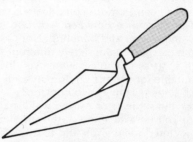

Walks, Walls & Patio Floors

By the Editors of Sunset Books and Sunset Magazine

LANE PUBLISHING CO. • MENLO PARK, CALIFORNIA

Foreword

The decision to add a patio, lay a walk, or erect a wall is not a difficult one. But after that decision is made, the matter becomes more complicated: what to build with, and how? You'll find a world of difference between redwood chips and concrete surfacing — and not merely in the weight and handling. Not only do they vary substantially in cost but also the two serve completely different purposes.

Walls also present you with an array of choices. Once you locate a spot for a wall, should it be a low brick wall suitable for sitting? Or do you want the privacy provided by tall concrete block screen? How do you decide?

This book is concerned with the subjects of garden walls, patio paving, and step building. It is designed to help you unravel some of the problems you encounter in deciding which material is best suited for the job and to guide you in the actual construction. Since almost all garden masonry projects involve a substantial investment of both time and money, careful planning is essential.

The book is organized according to construction materials; for each you'll find some ideas for use, followed by suggestions for actual construction techniques. When a basic technique is the same for two or more materials, it will be discussed once in the chapter where it applies most logically and then cross-referred to in the other chapters.

Many of the ideas in this book have been featured in Sunset Magazine. Also, we wish to thank Alex Grossi, Pacific Ready-Mix, Inc.; Ted Holmes, Freeman-Sondgroth Construction Co.; William Louis Kapranos, landscape architect; National Concrete Masonry Association; Portland Cement Association; Hans Sumpf, adobe manufacturer; J. G. Torres, concrete contractor; and Pete Wismann, masonry contractor, for their help in compiling and verifying technical information for this book.

Edited by René Klein

Design: John Flack

Illustrations: Joe Seney

Cover: Photograph by Ells Marugg

Editor, Sunset Books: David E. Clark

Fifteenth printing December 1984

CONTENTS

Walks & Patio Floors

Nothing can dull the sparkle of owning a new home faster than that first walk into the back yard. In most cases, the new owner steps immediately into a forbidding wasteland complete with rock-hard soil, jagged chunks of leftover concrete, and a vast assortment of deep-rooted, stubborn weeds. For the new-home owner, changing this scene will probably become a first priority task. Even if you have lived in your home for some time or have just bought an older home, an overhaul of the yard might be necessary to make it better suit your needs.

Garden construction often means digging up the ground, and therefore should come first—before the rest of the landscaping is put in.

In any case, the plans will more than likely include a patio, walks, or both. You increase the living space of the house when you add a patio. It can be an outdoor living room, a private terrace off the bedroom, or provide an outdoor dining room and cooking area. Make it large and expansive, adjoining a number of rooms, or fashion it into two or more separate, smaller, more intimate patios. The combinations and functions are many.

Much more than a mere passage from one area to the next, a walk or garden path is a versatile landscaping tool. A walk divides the garden, providing a natural break in the landscape. Use it as a border for flower beds or a lawn or curve it and lead your guests through the garden the long way. A curved path is an excellent way to break the straight lines of a rectangular yard.

Once you establish the location of the patio and walks, you will have to decide what paving material to use.

This could be the toughest part of the entire operation; you can liken it to shopping for a new car. Do you want the traditional versatility of brick or the rustic old styling of adobe block? Is the temporary economy of loose aggregate desirable, or do you prefer the permanence and durability of concrete? Do you like the dark dull finish of asphalt or the slick polished look of tile? Will the classic look of flagstones or the natural look of wooden rounds and blocks best complement the house?

This decision will take some thinking. Don't be too hasty, because once you've laid paving materials, they're likely to remain in place a long time.

Basic considerations

Probably no one type of paving will completely meet all of your specifications. Each has its strong and weak points that you will want to assess. As you read through the following chapters discussing the various building materials, keep in mind the following considerations:

Surface texture. How will the paving be used? Will it need to be smooth for children to play on, or will it need to be rough to provide sure footing for the senior members of the family? Will it be slick enough for dancing, yet rugged enough for action games? Will it need to resist stains around the barbecue area? Are you going to be able to move furniture around without marking the surface too badly?

Appearance. Are the color and texture and pattern going to blend with your house design and garden plan? Do they have to match the indoor flooring? Will they reflect the light on the shady side of the house or will they glare too much in the sun?

Maintenance. Can the paving be cleaned? How readily will it show dirt and dust? Can grass clippings and leaves be removed? Will weeds grow through it?

Durability. Do you want the paving to be permanent or temporary; are you willing to work it over each year? How will it stand up under your local climate conditions? What will be the effect on it of snow, water, frost, or extensive heat?

Cost. How does the cost of the material stack up to your budget? Have you considered possible hidden costs of "sleepers" such as drainage provisions, unstable soil, or special construction processes?

Application. Can the materials be delivered to the point of application? When is the best time to install them? Will it be a one-man operation or are you going to need help? How long is the job going to take and what inconveniences might you have to put up with? How soon after the application can the surface be used?

First things first

Regardless of the type of paving you choose, you will, in most cases, have to prepare some sort of foundation or sub-base that will affect the life span of the material. The foundation almost always determines the appearance of the finished job; laying it probably contributes to the majority of the blisters on your hands.

If you enjoy an even climate and live on stable and well-drained soil, you can sometimes dispense with a foundation. But if the soil is shifty and the climate fluctuates between extremes of hot and cold, the paving should be protected with a pad of rock or gravel. This pad keeps water from collecting directly under the paving and protects it from the effects of the soil heaving because of moisture locked in the ground.

Drainage. The success and durability of any garden paving rests largely with the stability of the ground below it; good drainage is a major ingredient. Often, a rock pad or bed of sand provides adequate drainage, but sometimes additional provisions are necessary.

Special perforated plastic drain pipe placed in a narrow trench (about 12 inches deep, under the center or around the edge of the paved area) will draw off most of the water that collects under the paving. Place the pipe in the bottom of the trench, perforated side down. Pack gravel around the pipe to a depth of six inches and replace the soil above it. If the trench is placed below the paving, keep the fill soaked for a few days to be certain

COVER PHOTO shows pleasing brick-on-sand walk; bricks laid jack-on-jack between wooden headers.

that it is properly packed. For severe drainage problems, consult an expert before paving.

Grading. Once you're satisfied with the drainage, you can grade the site to the right depth for the particular paving, as recommended in the chapters to follow.

If you have to dig out soil, disturb as little earth as possible. If you dig too deeply, tamp in fill, soak it, and tamp again after it settles. Filled areas take many months to reform into the firm texture of native soil.

To determine how deep the area needs to be dug out, use a line-level to locate high and low points and to establish grade.

First, stake off the area in squares, say 5 or 10 feet on a side. Next, determine the grade for the surface of the finished paving and mark this point on a stake with a saw cut (chalk marks rub off). Attach a chalk line at the cut and stretch it down a row of stakes. Hang the line-level in its center, drawing the line taut. Level the string and mark the point where it touches each stake with a saw cut. Repeat with other rows of stakes until all are notched at grade. You now can find out how far down to dig: on each stake, measure down the thickness of the paving, plus its subgrade, plus an allowance for pitch of 1 inch in 10 feet.

BRICK PAVING

Motoring through Holland, tourists frequently travel over friendly, neatly patterned, brick-paved streets. Dutch road builders found bricks easy to handle and place, easy to maintain, and long lasting. For the same reasons, brick paving is effective in the garden.

Paving with brick: pro and con

One of the greatest pluses of using bricks is the compact size of the unit you are dealing with. Bricks can be handled and placed with relative ease. And cutting them to size for fit is a simple operation. As with most types of paving, of course, you'll need to exert considerable effort in preparing the site before the surface can be applied. Grading and screeding are time-consuming operations.

The uniform size of the bricks makes planning and estimating the quantity needed for the job fairly simple. Because a brick is a relatively small building unit, it will probably take 2000 of them to pave an average size patio.

Under most conditions, bricks produce a solid and durable paving surface. A hard winter can cause a brick on sand surface to buckle or warp, but this can be easily ironed out in spring by picking up a few bricks and filling in where needed with sand.

The rougher surface of the variety of bricks usually preferred for garden paving reduces glare and provides good traction. Because the surface is porous, bricks readily soak up water. As it evaporates, it cools the air and makes the surface cool under foot. Keep in mind that bricks can also absorb oil, grease, paint, and the like—all of which could leave stubborn stains.

Brick is a good paving choice because it is available in a wide variety of colors and textures. You'll find a brick to complement almost any kind of garden situation, as the following section explains.

Choose from a wide variety

Though the basic form and composition of brick has remained unchanged for almost 5,000 years, today's builder can choose from at least 10,000 different combinations of colors, textures, and shapes. Of this bewildering variety turned out by the brick yards, two basic kinds are favored for garden paving: 1) slick surfaced face brick and 2) rough textured common brick.

Most garden paving is done with common brick. People like its familiar color and texture, and it has the undoubted advantage of being less expensive than face brick. Common bricks are more porous than face bricks and less uniform in size and color (they may vary as much as 1/4 inch in length). Face bricks are not as widely available as common bricks; you'll notice that they are used more frequently for facing walls and buildings than for paving.

Three types of the more popular common brick can be ordered:

Wire-cut brick. This brick is square-cut and has a rough texture with little pit marks on its face. Lay it to expose the edge if you want a smooth surface.

Sand-mold brick. Slightly larger on one side because it must be turned out of a mold, this brick is smooth-textured and easy to keep clean.

Clinker brick. This brick is noted for its black patches and surface irregularities caused by over-burning. It gives a rough cobblestone effect.

Try to buy common brick that is hard-burned; it will outlast the softer "green" brick. When well-burned, brick is usually dark red; under-burned brick is salmon colored. Give the brick the "ring test": if you strike a well-burned brick with a hard object, it will give a clear, high-pitched metallic sound; an under-done brick, though, will respond with a thud. Because the softer brick wears unevenly, it can be useful if a weathered look is desired.

Used bricks, with their uneven surface and streaks of old mortar, make an attractive informal pavement. Taken from old buildings and walls, these bricks are usually in short supply. Many manufacturers are now converting new bricks to used bricks by chipping them and splashing them with mortar and paint. Manufactured used bricks cost about the same as the genuine used ones and are easier to find.

The exact dimensions of a standard brick vary from one region to another and from manufacturer to manufacturer, but a brick generally measures 2¼ by 4 by 8 inches. Today you can find many units larger and smaller than this that are excellent for paving. Such bricks belong to the "split" variety and are roughly half the thickness of standard bricks.

How to order

Once you have chosen the right brick for the job, shop around to see where you can get the best bargain. Prices can vary from area to area. When you order, ask about delivery charges. They are usually low but often not included in the quoted price. It's also a good idea to pay a little more to have bricks delivered on a pallet. This prevents what can be considerable breakage in unloading when the bricks are merely dumped off a truck.

Make sure your dealer has sufficient bricks of the variety you need to see you through the entire project. If you have to complete the job with a different variety or some from another supplier, you may find it impossible to complete the lineal or color pattern that you have started, especially if you are fitting bricks closely together.

BRICK VARIETIES cover wide range in colors and textures. Choices vary from light-colored, hard-fired buff bricks (above) to rustic, rough-textured, chipped and stained used brick (left).

ENTRYWAY PAVED with bricks in mortar; circular pattern consists of reversed panels of running bond.

RAILROAD TIES serve dual purpose as header, framing bricks, and as garden steps where level changes.

EXPANSIVE BRICK PATIO is uninterrupted by headers; jack-on-jack paving pattern minimizes brick cutting.

ALL-BRICK PATIO features basket weave pattern of bricks set in mortar divided by wooden headers.

BRICK TERRACE laid running bond on sand without headers; brick pattern is well suited for large areas.

RUSTIC USED BRICKS in sand, framed in 2 by 4 headers, make broad, decorative garden walkway.

RECTANGLES OF BRICK laid in wood frames below surface protect grass from foot traffic.

BRICK PATH set in half basket weave; wooden headers were placed so that no bricks needed cutting.

BRICKS SET IN MORTAR form durable stepping stones and add decorative design to dichondra lawn.

DIAGONAL HERRINGBONE brick pattern separated from running bond by concrete divider for variety.

VARIETY IN BRICK PAVING achieved by combining three brick patterns; design requires brick cutting.

USED BRICKS were paired with exposed aggregate concrete surface to add variety to patio paving.

How to Lay Brick

Before you begin, consider these preliminaries: what pattern or combination of paving patterns will you use? What is involved in preparing the base? You can choose from a number of basic patterns or try your hand at creating your own; cut out some cardboard bricks and give it a try. Since preparing the base is a crucial task, it deserves your best planning. No matter how carefully you set the bricks, if the base shifts or settles, you may have to start all over again.

Your choice of patterns

In choosing the pattern (also known as the bond), keep in mind the degree of difficulty involved.

Some bonds demand a good deal of accuracy and brick cutting. Your choice of bond will be affected by whether you lay the bricks with closed joints (butted together) or open joints (spaced).

Dimensional variations of common clay brick make some patterns difficult to complete when the bricks are laid lightly butted. A pattern like jack-on-jack, for example, is hard to follow through a large area. When laid solid, basket weave will produce a curious effect: a ½-inch hole turns up in the center of each block of eight bricks. If you are using a common brick that has variations in brick dimensions, an open joint gives you the flexibility to take up the difference in the brick sizes.

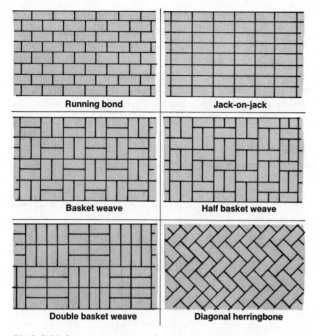

CHOOSE from a variety of paving patterns. These are six of the most popular; use open or closed joints.

The six patterns shown above are the most popular; the top two are the easiest to lay.

If you are paving a large area with a single bond, the surface may become monotonous. This problem can be eliminated in a number of ways. Try changing the direction of the bond to add variety to the paving. Another method is to combine two patterns. For example, you can break up a large surface of diagonal herringbone by dissecting it with a few rows of the jack-on-jack bond. Or break up a basket weave pattern with a double row of the running bond.

A very effective way to add variety to brick paving is to introduce a different building material, such as wood or concrete. Break up a bas-

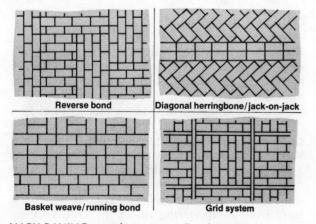

VARY PAVING over large area. Combine two or more brick patterns or build wood or concrete grid.

ket weave pattern with a grid system of redwood or cedar (the building technique is the same as for headers or wood edgings: see page 21). The same grid system idea can be created with strips of concrete (see the chapter on concrete paving, page 22). If you decide to use the concrete grid, plan your sections carefully to be sure of a good fit.

Gaining in popularity, railroad ties have been used very effectively recently in conjunction with brick paving.

Don't be afraid to use your imagination to add a little personality to your brick work. Here is your chance to shine—but take care not to get carried away and make the pattern too busy, especially if you are paving a smaller area.

Preparing the base

A major part of successful brick laying is in the preparation of the base.

The ground should be solid. If you have to lay the bricks on fill, be sure it has settled for some time, preferably for at least a year. The fill should always be wetted and tamped or rolled before you lay the bricks.

The extent of drainage that you'll need depends upon the soil and how the bricks are to be laid. Bricks on sand will drain well if the subsoil drainage is good. If not, the paving must be sloped (as for bricks in mortar) to allow surface drainage. Another approach is to install drainage underground (as mentioned in the introduction of this chapter). If drainage is unusually poor, consider seeking professional help to correct the problem.

Make a careful study of aggressive plant root systems, especially if you're planning to lay bricks in mortar, or plan to use a grid system such as those mentioned above. Exceptionally vigorous roots can make the best-laid patios go awry.

Laying bricks on sand

A patio of brick on sand is one of the easiest paving projects for the beginner. Unless you live in an area where the ground freezes, bricks on sand provide as durable a surface as bricks set in mortar and—contrary to what you might expect—one that is as permanent as you want it to be. An added advantage: if you decide later that the patio needs a change, you need destroy only one brick to get the rest out in perfect condition.

Grading. Lay out the base grade with a line and level. Mark the heights with stakes at the edge of the area to be paved. Stretch a line tight between the stakes and measure down from the line to allow for the thickness of the brick and the base.

It helps to loosen up the top soil so that you can cut and fill with soil as needed to establish a relatively even surface. The surface should be pitched slightly (roughly 1 inch in 6 to 10 feet) for proper drainage. Tamp the soil well. Sand will take care of slight irregularities in the screeding (leveling) process.

Edging. Give a firm edge to this kind of paving both for appearance and for permanent stability. For ideas and techniques for construction of edgings (or headers as they are often called), see the box on page 21. Once the perimeter is established by being outlined with header boards, you can set up movable header boards (to be moved along to accommodate the screed, see below) or a permanent grid system. Plan to lay a sample section before installing the grid permanently in order to check the fit and minimize brick cutting.

Screeding. An important operation, screeding is the actual leveling of sand to a uniform surface on which the bricks will be placed. The screed is normally a 2 by 4 with a 2½-inch extension nailed along the bottom, two inches or so shorter than the 2 by 4 on both ends. It rides on the headers, leveling the sand as it is pulled along. Keep the length of the screed down to a manageable size—say 4 to 6 feet, or short enough to fit in the grid system if you have installed one.

Screed about 3 feet at a time. This will provide a base of about 15 square feet of brick paving. As you proceed, you will be moving the temporary headers along, keeping them spaced equal to the length of the extension of the screed. Spread about 1½ inches of sand inside the headers and wet it down to help it settle. Level it by pulling the screed towards you so that the sand is

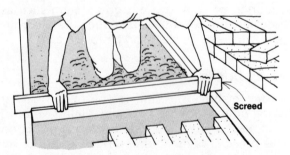

Screed

PULL SCREED along temporary headers to level sand, provide uniform bed on which to lay bricks.

about 2½ inches below the headers. For a good firm base, tamp the sand after screeding; add more sand, if needed, and screed again. Don't walk on the base after the final screeding.

Setting bricks. Lay the bricks in the pattern you have chosen, tapping each one into place (pref-

erably with a rubber mallet) and working it down level. Since the bricks will settle, it's a good idea to set them a little high. A helpful hint: start working from the middle of one side to lay a ribbon of your pattern all the way across the area in whatever direction you wish it to go. Then work from this middle ribbon towards each side. As you get near the edges, tighten or loosen the joints as necessary so that you end up with whole bricks whenever possible. Check the pitch, for drainage, as you go along.

When the bricks are down to your satisfaction, fill in the joints with fine sand, such as a 30-mesh. Throw the sand out over the top of the brick surface and let dry for a few hours, then sweep it into the cracks. Repeat this process until the joints are filled; then wet the area with a light spray so that the sand in the joints will settle completely.

The most common way to set brick on sand is with closed joints (butted snugly together). Bricks on sand can be set with open ½-inch joints, but this surface is not as stable as the former. Bricks laid with open joints can work themselves loose as the sand settles, possibly requiring you to refill the joints from time to time. But for common brick that varies considerably in size, open joints may be the only answer.

How much to buy. The amount of material to buy will depend on whether you use open or closed joints. *For a 100-square-foot surface bedded on a 1-inch bed of sand, order the following approximate quantities:*

	Closed joints	½-in. open joints
Bricks	500	450
Sand (in cubic feet)	9	13

Weed control. Some weeds will inevitably straggle through brick-on-sand paving. Periodically, pour a general contact weed killer between the bricks.

Bricks in dry mortar

If you like the idea of the open joints in brick work but shudder at the idea of having to rework the patio from time to time, use the dry mortar method.

Set the bricks in the same way as bricks on sand with open joints but then add portland cement to sand that is to be swept into the joints. Wet the whole paving down with a fine spray. As a preliminary option, add cement (1 to 6 ratio) to the sand bed; this will help keep the foundation from fraying at the edges and will stop the sand from disappearing down the cracks in the ground. The mortar in the joints prevents weeds from coming up between the bricks and keeps

TOOLS OF THE TRADE

To join the home mason's ranks, you will need a few tools. A quick search of the tool shed or work shop will usually turn up a good number of them: a hammer (illustration 1) to build your edgings and forms; a good saw (4); a 2-foot spirit level (5) to check horizontal plane; a plumb bob or makeshift plumb bob (10) to check a wall's true vertical plane; a carpenter's square (9) to keep your corners true; a measuring device (12 and 15) such as a retractable steel tape, a wooden zig zag rule, or a linen measuring tape; some string or a chalk line (17) to help guide along a straight and true line; a shovel (6 and 7) to dig a trench for the foundation of a wall or to grade the soil before the paving goes down; a pick (16) to break well-settled stubborn ground; a hoe (13) to mix concrete or help spread it.

Many home masonry tasks will require some more specialized tools you may have to borrow or buy: for example, a pointed trowel with a 10-inch blade (2) used for buttering is a must for most jobs involving mortar. For concrete surfacing, you will need a wooden float (8), a bull float (14) to float the concrete, a rectangular steel fanning trowel (3) to finish the concrete surface, an edger (21) to form smooth edges, and a jointer (11) to cut control joints.

If the plans call for bricks, broken concrete, stone, or adobe block, you'll find a broad-bladed cold chisel or brick set (18) a handy tool for cutting and dressing the materials.

To fill out your mason's tool bag, you may want to invest in some optional paraphernalia: a couple of jointers (19) to shape mortar joints; a mason's hammer (20) to chip away rough edges of a cut brick, block or stone, and a tamper (22) to pack soil.

Rent any needed heavy equipment.

Some tools, such as a screed for leveling a soil or sand bed before the paving is put down, can be built on the spot.

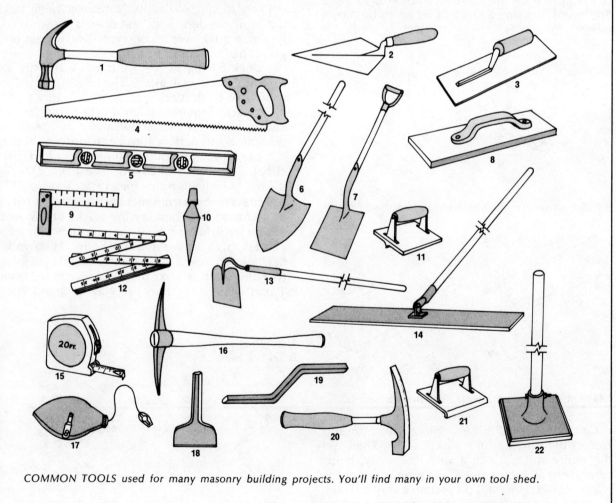

COMMON TOOLS used for many masonry building projects. You'll find many in your own tool shed.

water from washing away the foundation. This type of paving provides a good solid surface where the subsoil is sound and where severe frost is not a problem.

Caution: although dry mortar is not difficult to work with, some will usually stick to the surface of the bricks as you sweep it across (see below), leaving a mortar stain. This may not be a disadvantage if a rustic look is desired.

Laying the brick. After the 2-inch sand bed has been laid and screeded, lay the bricks on the sand, leaving a ½-inch joint between them. Continue this until all the bricks have been set in place. Mix the dry mortar, using 4 parts plastering sand and 1 part cement. Throw this mixture across the bricks, being certain that the bricks are dry before you scatter the mortar mixture across them to prevent mortar from sticking to the brick surface. Then sweep it into the cracks. For this operation, kneel on a long board to distribute your weight and avoid disturbing the bricks. With a hand brush, push the mixture between the bricks; then with a broom, sweep away the excess. You can tamp the dry mortar into the joints to get a firm bond by using a piece of wood ½ by 4 by 6 inches.

BRUSH dry mortar mixture into joints with stiff brush.

TAMP mortar into joints with small piece of wood.

Once the mortar mix is in place, clean the surface of the bricks as much as possible. Then wet the entire surface with a very fine spray, being careful not to splash any of the mortar out of the joints. Avoid doing this project on a very hot day, for the bricks will suck the water from the mortar

too fast, weakening the joint. On the other hand, take care not to wet the bricks so much that spill-over dilutes the mortar.

When the mortar has begun to harden, you can smooth (rake) the joints with an S-shaped jointer, wooden dowel, or broom handle to give the job a finished professional look.

Let the mortar set for about two hours, then scrub each brick again with a wet burlap sack. When the paving is completely dry, go over it with mortar cleaner (see the section on brick clean-up).

What to buy. *For 100 square feet of paving on a 2-inch sand bed with ½-inch joints, you will need approximately:*

Bricks	450
Cement (sacks)	1
Sand (cu. ft.)	17

Bricks in wet mortar

In localities where winters are severe or in an area where you are handicapped with unstable soil that is likely to slip and settle, a more durable foundation is needed. The best plan is to set bricks in wet mortar over a concrete slab poured on a gravel pad.

Brick paving in wet mortar is a difficult job for the home craftsman. The method used by professionals (buttering each brick before setting it in place) requires much practice, but there is another method that allows even the beginner, with time and patience, to get handsome results.

Lay a concrete foundation as described in the chapter on concrete paving (see page 22). Wet the bricks in the morning for an afternoon job. This will prevent them from sucking the water out of the mortar mixture. Then lay the bricks in a ½-inch mortar base, leaving ½-inch joints between them, and let them set for about 4 hours. Then with a flat, pointed trowel, push the mortar into the joints. (For this mortar mix, use 3 parts graded sand, 1 part cement, ½ part fire clay or lime.) When

PACK wet mortar into joints with pointed trowel.

FINISH mortar joints with jointer or wood dowel.

the joints have hardened for about ½ hour, rake them with a jointer or other rounded tool as shown above.

What to buy. *For 100 square feet of paving with ½-inch mortar joints (not including the concrete base), you will need these approximate materials:*

Bricks	450
Cement (sacks)	2
Sand (cu. ft.)	12

Bricks over existing concrete

If you have a concrete patio or path already installed that you'd like to replace, it is not difficult to pave it over with bricks. For this type of project, brick does not need thickness for stability. A variety of split brick (half the normal thickness) will do the job.

After setting up headers around the concrete, lay the bricks, using either the dry mortar or wet mortar method described earlier.

Cleaning the brick

The first clean-up problem you will encounter is usually that of spilled mortar. When you drop excess mortar on the bricks, you can save time and effort by wiping it up immediately with a wet piece of burlap. If allowed to dry, mortar can leave nasty, stubborn stains. To remove these stains, soak the area to be cleaned, allowing the water to soak in. Then, wearing rubber gloves and using a plastic bucket, brush the area with a 9 part water to 1 part muriatic acid solution. Rinse thoroughly with water to prevent acid burn stains from remaining on the brick.

Efflorescence. The white deposit that frequently appears on the brick paving after rain or hosing is caused by water-soluble salts in the bricks or in the mortar bed rising to the surface. When the water dries, salts appear in the form of a white crystalline crust.

Normally, this condition recurs for several years until the salts have all come to the surface. Until then, you can use a hard brush to brush the white crystalline deposits away as they appear. Don't try to hose them away; the salts will simply dissolve and soak back into the paving, later working their way up to the surface again. In extreme cases, you may find it necessary to use hot water and detergent.

Most experts recommend that you not try to combat efflorescence by applying a sealer to the paving (or walls) since deposits can build up and appear beneath the surface sealer, where they will be very difficult to remove.

Though it is difficult to do, the best deterrent to efflorescence is to keep water from penetrating beneath the patio into its foundation. One way this can be accomplished is to put a plastic film beneath a setting bed or concrete under-slab. Providing good surface drainage will also help minimize the problem.

HOW TO CUT A BRICK

No matter how carefully you plan the work, some brick cutting is almost inevitable, especially when you are setting a bond where the bricks overlap, such as the running bond or diagonal herringbone patterns. Save your cutting of bricks to the last so that you can do them all at one time and be more certain of the size and the shape.

The best tool to use for the job is the brick set (broad chisel). Set the brick on flat sand and place the chisel (with the bevel facing away from the end to be used) along the desired cut. A hard tap with a heavy soft-headed steel hammer will cut the brick (wear safety goggles). For cleaner results, tap the brick lightly to cut a groove across all four sides before giving it the final severing blow. If needed, chip the rough edges with the chisel or mason's hammer.

When your bricks are delivered, a few are bound to be broken; save these for cutting.

This same cutting technique can be applied to a number of other masonry products, such as concrete blocks, flagstones, and adobe blocks.

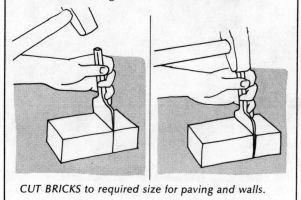

CUT BRICKS to required size for paving and walls.

ADOBE PAVING

Historically, adobe structures were doomed to eventual decay, falling victim to the annual duel between summer heat and winter rain. Today's adobe, however, has little in common with its 18th century predecessor because it carries an asphaltic stabilizer that keeps the adobe bricks from dissolving.

Paving with adobe: pro and con

Few paving materials can add rustic tone and historic elegance to a garden living area the way adobe does. Adobe paving gives a warm and friendly air to an informal patio. Spaced with 1-inch open joints, adobe provides an excellent base for a living floor of creeper and moss that grow to fill the joints and soften the look of the surface.

If you live fairly near the source of supply, you will find that these large blocks can cost less per square foot than brick. Even though adobe is found almost entirely in the Pacific Southwest, primarily because the manufacturing facilities are located there, it can be used effectively in any section of the country. If you live outside the West, though, delivery charges can add considerably to the cost, perhaps making adobe an un-

economical paving choice for you.

Your choices of adobe

All adobe bricks commonly used in construction are made by the same process and have nearly the same textural quality. Blocks are available in a variety of sizes; the most commonly used is the 4 x 16-inch face size with variations in depth ranging from 3½ to 12 inches. In addition, a non-standard block is designed specifically for paving use. This block, available in face sizes of 12 x 12 inches or 6 by 12 inches with a 2½-inch depth, contains more asphalt than can be used for blocks in walls. The special size of this block prevents it from being mixed into wall work. But all sizes can be put to garden uses. Blocks vary in weight from 12 to 45 pounds.

Laying adobe blocks

Adobe blocks are laid on a sand bed in about the same manner as clay bricks. The sand bed must be solid, stable, and quite level. If adobe has to straddle a hump or bridge a hollow, it could crack when weight is put upon it.

LARGE ADOBE BLOCKS laid jack-on-jack with open joints in patio of SUNSET'S California home office.

STURDY EDGING helps keep large blocks in place; open joints take up size irregularities of adobe blocks.

LAY BLOCKS with open joints to allow for size irregularities; fill with sand, soil, or dry mortar.

Because dimensions of adobe blocks may vary slightly, it is usually difficult to lay them in patterns that call for tight fitting. One of the best patterns for laying adobe is the jack-on-jack bond. Using this pattern, lay the bricks with either open or closed joints as with clay bricks (see page 10). Open joints may be filled with sand or dirt. The blocks are heavy enough to stay put with sand-filled joints; filling with dirt permits crevice planting that adds to the rustic appearance of adobe paving. Laying the blocks in sand or dirt allows for good drainage and helps to extend the life of the blocks.

Joints between the blocks may be filled with dry mortar, but this method should be used only when it is absolutely essential. Adobe manufacturers have found that adobe failure, aside from normal wear, frequently starts at the mortar joints because of the salt action during wetting and drying in the winter months. If you do wish to use mortar, use 1 part cement to 3 parts sand mixture. Asphalt stabilizer may be added to the mortar mixture (1½ gallons per sack of cement) as a waterseal.

Cutting adobe blocks. Using the jack-on-jack pattern will make planning for the number of blocks you'll need a simpler task. And it will almost eliminate having to cut the blocks. But adobe is not difficult to cut in any case. Use either a hatchet or a brick set (see page 15).

How to order

Most adobe brick is sold through dealers; check your telephone book. Dealers who handle adobe usually carry only a small stock in their yard but can arrange for larger shipments directly to the site. Make certain you inquire about the delivery arrangements.

How much to buy. *For 100 square feet of adobe paving set with 1-inch joints, you will need approximately these supplies:*

Adobe blocks (average size)	100-120
Sand (in cubic feet)	
For sand joints	11
For mortar joints	9
Cement (sacks)	3
Stabilizer (gallons)	5

Although these estimates are intended to give a general idea of the amounts and materials needed, exact figures will depend upon the size of the blocks chosen. In any case, it is always a good idea to buy a dozen extra adobe blocks for replacements. In a large quantity of the blocks, a few are likely to come with irregularities or will develop flaws after you put them down. You may have difficulty replacing them with blocks that exactly match the color and texture of the first batch.

SQUARE ADOBE BLOCKS designed specifically for paving adds friendly, spacious feeling to patio area.

GARDEN WORK CENTER paved with adobe blocks; slanted surface affords good drainage and is fast drying.

LOOSE AGGREGATE

Loose aggregate is a low-cost, easily-applied material that can be used to supplement existing paving. With it, you can pave secondary areas, such as a path, service yard, or potting shed. It is excellent as a temporary stopgap to make a muddy area navigable through a winter or to rough-in a patio while the budget is convalescing from the shock of buying the house.

An aggregate does have some inherent weaknesses as a paving material. It is a slow and uncomfortable surface to walk on; it allows weeds to grow through it; it must be replenished from time to time; and, if used next to a lawn, it tends to get kicked into the grass where it can nick the blades of a lawn mower or be hurled about at bullet speed by power mowers.

Choose from a variety

For whatever purpose you may need loose aggregate, you can locate a variety of kinds. Though low in cost, some are attractive and dressy, well worth renewing from time to time. Others are desirable because they fulfill a functional need.

Wood chips. Most wood chip products are by-products of mills and make serviceable paving for certain garden areas. Confined between headers (see page 21), they make soft, springy, and attractive paths. In addition, the grid system of headers will limit the tendency of chips to scatter. Wood chips are generally too easily scattered to be satisfactory for use in a large patio area.

Wood chips make an excellent protective cushion under swings and slides in children's play yards; they are gentle with small elbows and knees. Cost depends upon proximity to the site and the time of the year. They are difficult to obtain after logging has stopped for the winter.

Firbark. Warm, dark tones make firbark an attractive cover when spread loose in an area where there is little traffic or when used with headers for a path. Expect it to stay fresh for about three years with a minimum of maintenance. This paving requires good drainage because, like most wood products, it is susceptible to insect or bacteria infestation. Sterilize the soil before laying the

bark to minimize weed growth (fallen leaves can be left to disintegrate).

For 100 square feet of 2-inch surfacing, you will need 1 cubic yard of the standard ¾ to 1-inch mesh. It is also available up to 3 inches in larger mesh.

Fine-ground bark. About the texture of coarse sawdust, fine-ground bark is used primarily as a soil conditioner but can serve in play yards.

Mill chips. Made from mill scraps, mill chips make an attractive walk but splinter easily, making them unsuitable for play yards.

Sawdust. Rarely an eye pleaser, sawdust can be useful as a cushion under swings or as a sponge in muddy areas.

Tanbark. Gardeners consider tanbark a good landscaping material. Tanbark supplies are dwindling in some areas.

Gravel. Gravel, or crushed rock, provides a fairly durable surface for walks and, under quiet conditions, for patio floors. It stands up best when put down as a topping over a more permanent bed of redrock or decomposed granite. Put directly on solid earth, it will give good service for several seasons if renewed from time to time.

Advantages of gravel: a path of gravel or crushed rock dries quickly, and spilled dirt is easily washed away. Gravel also works well in combination with other paving materials, making an excellent base for stepping stones. And a band of gravel around tile or concrete paving absorbs the water swept off the solid surface.

To lay gravel, place it between headers of wood, brick, or concrete (see page 21). Rake it over the area in thin layers, dampen it, and roll it down. Repeat until the area is built up to final thickness.

For 100 square feet of 2-inch surfacing, you will need ⅔ cubic yard.

Redrock. Available under a number of other names (check your local dealer to see if it is obtainable in your area), redrock is a rocky clay that compacts solidly when dampened and rolled.

In garden paving, redrock can either be laid by itself or as a pad for a dressier topping. When

VARIED COLORS of gravel and crushed stone, separated by wooden edgings, make decorative garden paving.

WOOD CHIPS used to pave children's play yard; chips provide soft, springy cushion beneath play gear.

GRAVEL-PAVED PATIO confined by heavy wooden edging. Surface is fairly durable under quiet conditions.

CONCRETE STEPPING STONES make gravel patio area more navigable; place stones on (or recessed in) gravel.

GARDEN OF RAISED BEDS paved with gravel; paving provides excellent drainage and dries quickly.

it is put down alone, redrock provides a clean and hard surface. In time, the surface does wear away and break down into dust. When the surface is no longer satisfactory as a paving material, it can be used as a foundation for a more durable and permanent surface.

Put the redrock down in 1-inch layers, rolling each layer thoroughly until you have a 3-inch depth. Dampen each layer before rolling it.

You will need roughly 1 cubic yard for 100 square feet of 3-inch-thick paving.

Decomposed granite. Similar to redrock, decomposed granite wears better and costs about a third more than redrock. Lay it in the same way that you lay redrock.

Colored rock. Not as economical as ordinary gravels, colored rock can cost 5 to 10 times more. This variety of aggregate can be naturally colored —such as the pink and red rocks that are volcanic in origin—or it can be artificially colored. Take care not to use a color that will clash with your plantings.

Some colored rocks are available only in the area where they are found.

Dolomite. A less popular paving, dolomite is a clean limestone that requires extra care to keep its stark whiteness from discoloring. Though not expensive, dolomite should be used in small areas because of its coloration.

Its standard size of ¼ to ½ inch is small enough to make an impervious bedding advisable. Plastic sheeting laid underneath is recommended because it will limit the material's contact with ground water; prolonged contact can yellow the stone.

Crushed brick. Crushed brick is one of the more expensive loose aggregates, costing ten times as much as redrock. But if applied lightly over a stable base, it can be put down for moderate cost. Unlike gravel or crushed rock, crushed brick fragments break up and wear down; they are not advisable for an area that gets heavy traffic.

Crushed brick's venetian red is too intense to spread over broad areas. But if used in moderation, it brings to the garden a fresh and lively color, adding a striking accent to plantings.

Surfacing with loose aggregates

Regardless of the type of aggregates you decide to pave with, a header or grid system will almost always be a necessity (see page 21 for construction details). When working with loose aggregates, you will usually require curved edgings either for practicality, design, or both. To form curved edgings, nail thin, flexible boards together, staggering the splices, until the laminated board is the desired thickness around the curve. Many lumber yards stock "resawn" boards (½ and 1 inch thick) for curved edgings. You can also bend plywood or saw kerf 2-inch-thick lumber (see page 29).

Once the edgings are in place, it is always a good idea to place heavy grade polyethylene plastic sheeting on the ground surface before aggregates are put down. The plastic offers an effective bedding and serves to hold down weeds that have a tendency to sprout up. Sheeting should be punctured every square foot or so to allow water to drain away.

Adding stepping stones. An excellent way to make a loose aggregate walk more stable is to add stepping stones. Stepping stones are especially desirable if the walk must bear a considerable amount of traffic or must accommodate the senior members of the family.

You can choose from a variety of materials that make suitable stepping stones, but keep in mind that they must be flat in order to be stable in a loose aggregate surface. Pre-cast concrete paving blocks with an exposed aggregate surface are best suited, but flat stones or redwood rounds can also be used quite effectively.

Place the stepping stones in the loose aggregate surface so that the top of the stones extends slightly above the aggregates. This will prevent the aggregates from slowly covering the stepping stones. Take care not to set them too high so they become an obstacle rather than an aid. And make sure that the stepping stones are stable and do not rock or tilt when stepped upon.

Maintenance

The degree of maintenance required for a loose aggregate surface will depend on your choice of material. All loose aggregate surfaces require replenishing from time to time, and most will need an occasional raking or cleaning.

Replenishing. Gravel and small rocks require replacing least often. Deterioration is not a problem, but they will get scattered over a period of time or may be forced through the polyethylene sheeting into the soil. With moderate use, gravel or rock aggregates should last two years without replenishing. Because wood chips tend to break down, they should probably be renewed once a year.

Cleaning. In most cases, fallen leaves and other organic debris can be left to deteriorate and disappear in the surfacing. Keep in mind, though, that this decomposed material will encourage weed growth. If this debris is excessive, it should be raked and cleared away before it decomposes.

HOW TO BUILD EDGINGS

Whether you lay it, hammer it, set it, or pour it in place, a patio floor, walkway, or path will almost always require some sort of edging, often referred to as a header. In addition to outlining the patio or walk surface, headers confine the surfacing material within a desired area, an important function especially when you are using loose aggregates, pouring concrete, or setting bricks in sand.

Headers make neat demarcations between the patio and the lawn, flower bed, and other planting areas. But they also work effectively as decorative divisions or grid systems within the patio paving. They can separate different colors of loose aggregates, give texture and design to poured concrete, and divide a variety of brick patterns.

Most commonly, edging or headers are made of wood, but masonry can also be used effectively to give permanent edging to any paving. For wooden headers, use heart redwood or cedar — both are naturally resistant to rot. Coating the wood with a wood preservative prior to installation will help prolong its life.

To prevent the lumber from splitting (particularly stakes), use galvanized box nails; they are similar to box nails, but thinner.

The most popular edgings are made of 2 by 4-inch lumber. If no bending for curves is required, you can also use 3 by 4 or 4 by 4 lumber for heavier edgings. The area to be paved should be graded before headers or edgings are installed.

To construct a masonry edge, you can make a complete border of bricks laid up with mortar joints (see page 14). Lay them on a thin bed of mortar or concrete, flat, on edge, or on end. Another way is to make an edging of concrete (see page 29) or to use a combination of bricks and concrete.

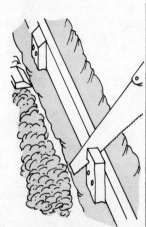

STRING guide line along outside area to be contained by edgings.

REMOVE soil carefully so that trench is slightly deeper than edging.

DRIVE stakes into ground no more than 5 feet apart, using heavy hammer.

NAIL stake to edgings; brace edging with 2 by 4 while hammering stake.

CUT stake tops on bevel with handsaw. Top of bevel should be as high as edging.

REPLACE soil on outside of edging; dig out inside soil to desired grade.

BRICKS set upright in bed of mortar make permanent edging.

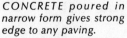

CONCRETE poured in narrow form gives strong edge to any paving.

CONCRETE PAVING

In garden construction, never has so much variety in textures and finishes been owed by so many builders to so few ingredients — only sand, cement, gravel and water. And because concrete is durable, the surface textures and finishes will last. If friends beat a constant path to your door, perhaps it is best they do it on concrete.

Paving with concrete: pro and con

Concrete is basically a mixture of cement and aggregates (sand and gravel or crushed stone) that hardens when water is added. It forms a paving that can be used to complement almost any garden scheme.

Concrete is permanent and very durable. If properly mixed and installed, it is down to stay with a minimum of maintenance. Concrete surfacing is not subject to decay and is relatively unaffected by severe climates and constant heavy traffic. And if you get tired of it, concrete paving makes an excellent foundation for bricks set in mortar (wet or dry, see pages 12 and 14).

Concrete offers a wide variety of surface textures and finishes to suit almost any need. Then, too, it's quite compatible with other surfacing materials. Use a strip of concrete to highlight a large brick surface or lay a few rows of bricks to break up an expansive concrete surface.

Concrete is an economical surfacing if you put it down yourself. The ingredients are available at reasonable prices almost anywhere in the country.

Concrete does have some disadvantages. In some situations, concrete can be a harsh and glaring surface. It must be mixed to exact specifications; there is little room for error. Once the ingredients are mixed and water added, you have to work fast; a mistake will require an extensive and perhaps costly redo. If it is carelessly installed or if drainage needs are ignored, the surface may buckle and crack.

Do it yourself?

When you decide on using concrete, you will probably wonder whether to attempt the job yourself or call in a contractor to do the work. The latter will take much of the initial economy out of using concrete; costs can double.

If you divide the job up into small sections and don't try to handle too large an area at once, you should be able to pour the concrete yourself with good results. But if you are in a hurry, have an unusually large area to cover, or face extraordinary soil or drainage problems, you would be wiser to call on a reliable contractor. First protect yourself and the contractor by setting up written specifications for the job (for suggestions, write the Portland Cement Association, Old Orchard Road, Skokie, Illinois 60076).

How to buy concrete

One of the pleasures of paving with concrete is that you can control the amount of personal involvement. You can choose to do as much of the job at any stage you like; it's all in how you buy the concrete:

Bulk dry materials. You buy the required amounts of sand, cement, and gravel and mix it yourself. This is the most economical method and allows the greatest amount of flexibility in how much you mix and pour at one time.

Dry ready-mix. You can buy sacks of all-purpose concrete suitable for most home uses (available in various size bags) containing correct proportions of sand, cement, and gravel, already mixed. All you add is water. Or buy a sand-gravel mix and add the cement and water. In this form there is no guesswork in fixing proportions. Buying dry, ready-mix concrete is usually the most costly way to buy concrete.

Ready or transit mix. Concrete in this form has been accurately mixed with water added at the plant and arrives at the site ready to pour in place. Some plants will deliver nothing less than a cubic yard or will charge extra delivery fees for anything less than five cubic yards. Check with the plant regarding their delivery and quantity requirements.

If you have ready-mix delivered, plan to bring the truck as near as possible to the site where you want the concrete placed. If the site is difficult to reach, be sure the driver knows this and ask him for suggestions before he attempts to deliver the concrete.

Some plants or building materials suppliers will rent small trailers that you can haul with your own truck or car; the usual load limit is about 1 cubic yard.

Whatever way you transport the cement, keep in mind that the site will have to be prepared to take all the mix that is delivered; you'll need to work fast. Without a doubt, helpers will come in handy at this stage.

Concrete formula

Mixing concrete can be varied to achieve two separate end results: 1) concrete without air and 2) air-entrained concrete. Your choice between these two will depend upon the local climatic conditions.

Concrete without air. *The formula recommended for garden paving is this:*

> 1 part cement
> 2½ parts sand
> 2¾ parts gravel or crushed stone
> ½ part water

The sand should be clean river sand; the gravel should not be larger than ⅓ the thickness of the slab (the formula above is based on 1½-inch aggregate); the water should be good enough to drink. Since the water-cement ratio is critical, it should be exact.

You'll find the easiest way to keep track of the formula is to apply it by the shovelful. For each shovel of cement, figure about 3 quarts of water (based on an average of 6 to 7 shovelfuls of cement per bag).

Air-entrained concrete. An air-entraining agent added to concrete mixture creates billions of microscopic bubbles in the concrete. These allow for contraction and expansion of water in concrete during a freezing and thawing cycle. Air-entrained concrete is highly recommended for cold climate regions. Adding the agent also makes the concrete mix more workable.

The amount of air-entraining agent needed in the concrete mix depends on the brand of the agent; consult your building materials supplier.

Follow the same formula for air-entrained concrete as you do for concrete without air with one exception: decrease the proportion of sand to 2½.

How much to buy

To determine how much material to buy, compute the square footage to be covered and refer to the table below. *It shows the amounts needed to fill 100 square feet:*

	Concrete thickness (in inches)		
Separate ingredients	3	4	6
Cement (sacks)	6	8	12
Sand (cubic feet)	13	18	26
Gravel (cubic feet)	18	24	36
Ready-mix (cubic yards)	1	1.3	1.9

EXPOSED AGGREGATE concrete is durable, colorful, non-slip surface. Wooden grid divides paving; outer edge is bricks set in mortar.

SMOOTH CONCRETE is excellent paving for back yard patio; it's easy to maintain, comfortable to walk on, and durable. Concrete poured in sections between 2 by 4 headers, floated and troweled.

SQUARES OF BRICK set in mortar add variety and texture to concrete floor of spacious back patio.

STREETSIDE PATIO of exposed aggregate concrete has one square left open for attractive planting.

SEPARATE CONCRETE PANELS connecting patio to deck were poured in place; forms were removed later.

CONCRETE POURED in metal ring set on circular earth mound later dug away to give pad floating look.

CIRCULAR CONCRETE SLABS set at different heights bridge pool and lead to circular sunken patio.

CIRCULAR CONCRETE TERRACE makes pleasing design in small garden. Curving path repeats circular pattern of concrete pad, also serves as border between flower beds and lawn.

SMOOTH CONCRETE slabs join deck and terrace; they're set below grass and don't obstruct lawn mower.

CONCRETE ROUNDS are set in lawn in random fashion; aggregate is white roofing gravel to add sparkle.

PAVING BLOCKS were cast in 4-module form. Four-block self-contained units make attractive walk.

CIRCLES OF CONCRETE seeded with colorful pebbles contrast with plants in friendly entryway.

PAVING BLOCKS set in gravel provide firm footing and interesting contrast on much-used path.

SQUARE STEPPING STONES set on 2-inch river rock lead from patio to lathhouse at back of garden.

JAPANESE BLACK PINE shaped to look like bonsai grows in open round left in concrete patio.

OPEN SQUARE and wooden grid system add decorative element to concrete patio. Curving floor adds depth.

NATURAL STONES and creeper form attractive border between curving concrete patio and lawn.

FREE-FORM planting bed in concrete swimming pool deck holds iris, succulents, rocks, breaks up paving.

MOSS COVERED ROCKS, crocus, dwarf conifer, and succulents in planting beds divide concrete paving.

CONCRETE PAVING 27

SHUFFLEBOARD COURT painted on smooth concrete, bordered with aggregate concrete, enhances patio.

BARBECUE PIT is built into round opening left in smooth concrete patio; border is bricks in mortar.

CONCRETE WALK curves at both ends. Remaining dirt-filled areas make pits for horseshoe court.

CONCRETE PAVING was poured around circular form; opening used for sandbox, later as planting bed.

How to Pave With Concrete

The most crucial parts of paving with concrete are careful planning and preparation. The paving process is of necessity a non-stop operation: once you begin pouring plastic concrete into a form, you can't stop until the form has been filled and the surface finished. Be sure your floor plans are definite as it is difficult and costly to alter them after the concrete has been poured. Make certain you have enough ingredients.

Preparing the site

Masonry paving requires a stable, well-drained foundation. This is even more true of concrete than it is of other masonry surfaces such as brick or tile. In addition, poured concrete needs sturdy forms or headers, either permanent or temporary, to confine the wet concrete within the desired boundary.

Grading. Stake out the area to be paved and remove grass, roots, and any other organic matter. Set the grade as described in the introduction to this section on paving (see page 5). A 4 to 6-inch gravel bed is recommended for areas where poor drainage is a problem.

After the forms are in place (below right), spread sand to a depth of 2 inches to fill in and use a tamper to create a uniform surface. For most home paving jobs, a hand tamper will do. Dampening the tamped surface before the concrete is poured in the form will prevent the fill from absorbing water from the concrete.

Building forms. Forms for concrete work are constructed in much the same way as the edgings or headers used for such other kinds of garden paving as bricks and loose aggregates (see page 21).

Because wet concrete exerts a lot of outward pressure as it is poured, the forms should be strong. For 3 to 4-inch slabs, use 1 by 4 or 2 by 4 lumber; for 5 to 6-inch slabs, 2 by 6 lumber is recommended. Wood stakes can be 1 by 2, 1 by 4, 2 by 2, or 2 by 4-inch lumber. Space the stakes no more than 4 feet apart. Driving the stakes slightly below the surface of the forms will make placing and finishing concrete easier. Use your 2-foot spirit level to be sure you are following the proper grade.

To curve the forms, bend plywood with a vertical grain or saw kerf 2-inch-thick lumber (cut ½ to ⅔ of the way through the wood). Wetting the lumber makes it easier to bend. Set the stakes closer together on curves (1 to 2 feet for short radius curves; 2 to 3 feet on long curves).

If the forms are going to remain as a permanent part of the pavement, use redwood or cedar that has been treated with a preservative. To protect wood from being marked or splintered during the finishing process and to prevent the wet concrete from staining the surfaces of the forms, mask the top surfaces with tape. Anchor the forms to the concrete by driving 16-penny galvanized nails

SECURE FORMS with wooden stakes driven into ground with top slightly below surface of the forms.

through the middle of the forms at 18-inch intervals. Do the same for interior divider strips, driving the nails from alternate sides of the boards.

Final form preparation. Recheck the level and grade by placing your spirit level on top of a long straight board placed on opposite ends of the forms. Make certain the fill is tamped and moist before the concrete is poured. The temporary forms should be dampened with water or coated with light-weight motor oil to prevent the concrete from sticking to them during removal. (Form release agents will also prevent sticking.)

CLEAR SITE of all plant growth, using straight-edge spade to cut and lift sections of grass.

Adding reinforcement. For large slabs, 8 square feet or larger, wire mesh reinforcement helps prevent cracking and shifting of the concrete. Check with your local building department for exact requirements. For most patios, ten-gauge mesh of 6 by 6-inch squares will do the job. Cut the mesh from the 5-foot roll to a size 2 inches smaller than the forms. For wider slabs, wire sections of the mesh together with a 6-inch overlap. Elevate the wire on pieces of masonry or wood about ⅓ to ½ way up from the base to the top of the form. Before pouring, make sure the mesh is flat and not buckled up above the form.

Mixing the concrete

If you are not using ready-mix concrete, you will have to mix the ingredients yourself at the site. Your choice of methods and facilities depends largely on how much mixing you have to do.

Hand mixing. The easiest way to mix by hand is to combine ingredients in a high-rimmed wheelbarrow; when the mix is ready it can be easily transported to the edge of the form. Other possibilities for mixing sites: build a wooden platform (6 or so feet square), use a square of plywood, or build a wooden boat (3 feet by 4 by 1).

When buying separate ingredients, the sand and gravel are bound to contain some moisture that will require you to vary the formula slightly. Plan to mix a trial batch of concrete to see what these variations will be.

Heap the ingredients onto the mixing surface or into the wheelbarrow, one shovelful at a time, keeping the proportions in line with the formula (see page 23). Spread 2½ shovelfuls of sand on the mixing surface and add 1 shovel of cement. Blend these two together and add 3 shovelfuls of gravel; blend until the gravel is evenly distributed. Scoop out the center and, using a pail marked off to show gallons and quarts, pour in 3 quarts of water and mix it with a rolling motion. Work from the inside of the heap outward. Adjust the formula by changing the amounts of sand and gravel added to the mix; keep the amount of water constant.

To be more accurate with the formula, construct a wooden box of one cubic foot to provide a more consistent volume.

Keep in mind that if air-entrained concrete is needed, the ingredients must be mixed by machine. Hand mixing simply isn't fast enough or vigorous enough to create the air bubbles in the concrete mix.

Machine mixing. For more ambitious jobs that require larger amounts of concrete, rent or borrow a portable mixer. These small machines, driven by gas or electric motors, generally vary in size from ½ to 6-cubic-foot capacity.

To mix the ingredients: shovel the entire required volume of gravel or crushed stone into the drum. Add half of the amount of water required while the drum is stopped. Then start the mixer and add the sand, cement, and the rest of the water. Allow this mixture to tumble in the rotating drum for at least 3 to 4 minutes.

The trial batch. Examine a sample of the first batch of concrete you mix, whether mixed by hand or machine. Work the sample with a trowel: the mixture should slide freely off the trowel but should not run off in a soupy dribble. You should be able to flatten the surface to a smooth compact finish, and the exposed aggregates around the edge of the sample should be evenly coated with the sand-cement paste.

If this sample does not measure up as a workable mixture, make some adjustments in the sand and gravel volume (don't tamper with water volume). Use the following guidelines:
● Too stiff and crumbly—reduce the amount of sand and gravel.
● Too wet and soupy—add sand and gravel to the mix. Make accurate note of the changes in the proportions and use the new formula for the following batches.

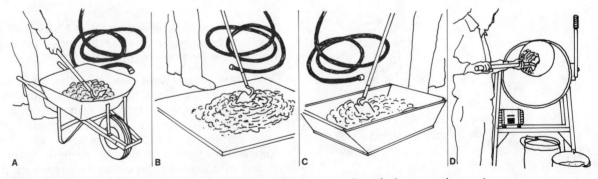

FOUR WAYS to mix concrete: A) in a wheelbarrow; B) on a wooden platform or plywood square; C) in a wooden boat 3 feet by 4 by 1; D) in a small power mixer you can buy or rent.

Pouring the concrete

Before you back the ready-mix truck or trailer up to the forms or start filling your wheelbarrow with your home-mix, stop to see if you have the necessary hands to handle the concrete. If you are using ready-mix, several willing neighbors are a must. If the ready-mix is poured directly into the forms, they can help direct and place the flowing mass of plastic concrete. If the truck or trailer cannot get close to the site, a couple of people will be needed to wheelbarrow the mix from the truck to the site.

POUR CONCRETE at one end of form and work toward other end. Spread with rake or shovel.

If you're mixing concrete at the site, it is helpful if one person continues the mixing while one or two others place and work the concrete.

Spreading. Pour the concrete starting from one end of the form. Work your way to the opposite end with subsequent batches. Use a rake or a shovel to move the concrete in order to spread it evenly. Don't over-agitate the mix because this brings too much water and fine cement "silt" to the surface.

Leveling. Before the form has been completely filled, begin leveling (also known as "strikeoff")

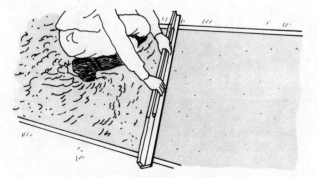

LEVEL CONCRETE with straight-edge placed across top of form, pulling toward you with zig zag motion.

the concrete from the same end where you began pouring. Placing a board across the forms, push the excess concrete along to the unfilled section of the form, using a zig zag, saw-like motion. Progress about a yard at a time, making certain low spots get filled. When the form has been almost completely leveled, fill the remaining forms with the desired amount of concrete.

Floating. Immediately after strikeoff, smooth the surface further by passing the flat surface of a bull float (for large surfaces) or a darby (for smaller surfaces within reach) over the wet concrete (see below). Push the float forward with the front edge slightly raised, then pull it back with the float flat against the concrete surface. This should fill remaining voids and level any rises. Fill in additional concrete where depressions persist and float again. When using the darby, keep it flat against the surface of the concrete and move it over the concrete in an arc with a zig zag, sawing motion.

BULL FLOAT (left) for floating large surface. Push forward with front edge raised; pull back flat.

DARBY (right) for floating small surface. Keep flat against surface, using zig zag motion.

Surface finishing

When you come to finishing concrete to its final desired surface texture, its virtues as a surfacing material become most evident. You have your choice of a number of finishes to suit your needs: a hard smooth finish, a rough wood float finish, a broom finish, or an exposed aggregate finish.

The first steps in finishing the concrete are edging and jointing. Edging compacts and smooths the edges along the forms; jointing provides a guide for and induces cracking along a desired indented course when the concrete expands and contracts during temperature changes.

Edging. Using a pointed mason trowel, cut the concrete away from the forms by running the trowel between the concrete and the form. Then push the edger along the opening between the concrete and the form, keeping the forward edge of the tool tilted slightly upward. Edge again lightly after the final finish has been applied except in the case of the exposed aggregate finish.

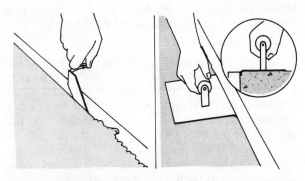

EDGE CONCRETE with edging tool; separate concrete from form with trowel; run edger along inside of form.

Jointing. Place a board with a good straight edge on top of the form (make sure it is lined up straight), using it as a guide for the jointing tool. Apply the same pressure and motion as with the edging tool. The resulting groove helps control the expansion and contraction of the material; permanent wood dividers accomplish the same end. Cracks that develop along these grooves will be concealed below the surface of the paving.

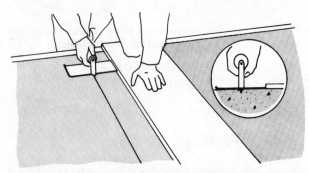

JOINTING TOOL run along straightedge cuts groove, helps control expansion and contraction of concrete.

Wood Floating. A wood float finish provides a surface that is smooth but not slick. Hold the float flat against the surface of the concrete and move it across in a sweeping arc with a slight zig zag motion. The resulting surface is fairly skid-resistant but more porous than the hard finish; it will soak up food stains more readily.

For a somewhat harder, smoother finish, use a metal float made of aluminum or magnesium. Work it in the same way as a wood float (see described process above).

WOOD FLOAT (left) gives smooth but textured finish. Steel trowel (right) makes surface hard and slick.

Troweling. To get a very dense, hard, and slick surface, use a steel trowel after you have smoothed the surface with the float (see above). For best results, trowel the surface twice, allowing a little time between to let the concrete harden a little more. The trowel is used in the same way as the float, flat against the concrete in a sweeping motion. Don't try this earlier than 30 or 45 minutes after the surface has received its rough finishing.

A troweled surface is quite glossy, easily cleaned, and can be waxed. But it is a highly reflective surface that may reflect the sun to an uncomfortable degree. This is an excellent finish for a patio floor that will be covered with a roof or for the floor of an area that will become an indoor room at a later time.

Broom finishing. An excellent way to get a textured, patterned finish is to broom freshly floated or troweled concrete. The prominence of the brush marks depends primarily upon the stiffness of the broom bristles.

To get this finish, pull the broom towards you, adjusting your downward pressure on the broom

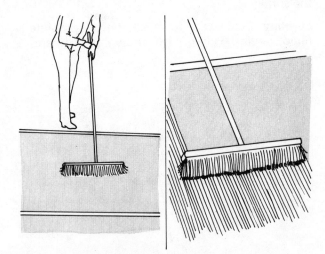

PULL BROOM across troweled concrete to give surface a textured and patterned finish.

to get desired results. The broomed texture can be straight, gently curving, or wavy.

Exposed aggregate. A very popular and colorful finish, exposed aggregate is one of the most time-consuming finishes to create. But you will be well rewarded for your extra efforts. The end product is a surface that blends well with almost any architecture and garden planting. Its pebbly texture takes the monotony out of a large expanse of paving and decreases glare.

Two of the most common ways of attaining an exposed aggregate finish are to expose the existing aggregate in the concrete mix or to seed aggregate in the concrete surface after the concrete has been put down.

To expose aggregate that has been mixed in with the initial mix, pour and finish the concrete through the floating stage. Take care not to over-float, for this will force the aggregates too deeply into the mix. When mixing concrete, use colorful aggregate, uniform in size. The ratio of aggregate to sand should be increased slightly over that in the regular concrete mix.

After the concrete has been poured and floated, wait until the water sheen disappears and the concrete has hardened to a point where it will support your weight when kneeling on knee boards without leaving an indentation.

HOSING AND SCRUBBING with broom exposes surface aggregates, gives colorful finish to concrete.

Hose the surface and gently brush or broom the sand and cement. If you can do this without dislodging the aggregate, you can continue the process for the rest of the paving. If the aggregate comes loose, wait until the concrete hardens a little more.

To seed aggregate in the concrete surface, pour concrete to a point slightly below the edge of the forms; strike off, bull-float, or darby the surface as before. Spread the aggregate, covering the entire surface. Using wood, a float, or darby, press the aggregate to just below the surface of the concrete. Float the surface, wait until the con-

crete can support your weight, brush off excess cement, and expose the aggregate in the same way as before.

Curing concrete

For optimum strength and durability, concrete must be kept moist while it is setting. It should not be allowed to dry out too fast by the sun or wind. This can be prevented by keeping the concrete wet or by retarding water evaporation. Wet down the finished concrete and cover it with burlap or canvas. Leave this covering on for about 10 days, keeping it damp for about the first 7 days.

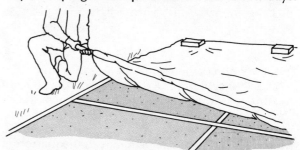

COVERING FINISHED CONCRETE with canvas or plastic for slow curing helps to give paving optimum strength.

A polyethylene cover can also be used. This plastic sheeting should be weighted down and kept flat against the surface to trap the moisture that would normally evaporate.

Special curing agents can be sprayed, rolled, or brushed on; they're available under a number of trade names. Preferred by many for their convenience, these agents are generally not as effective as the moisture curing methods.

Coloring concrete

Concrete can be very colorful—it is one of the easiest masonry products to tint. You can include the color in your plans and mix it in before you pour the concrete. Or you can add color as an afterthought, applying it after the surface has been put down. Use one of three basic methods to color concrete: 1) add color pigments to wet concrete before it is poured; 2) dust color pigment on the surface during the finishing process; or 3) apply color (paint, stain, or wax) to concrete paving after it has completely dried.

Mixing-in color. Mix the pigment with the cement and aggregates in a dry state first. Using white portland cement will produce brighter coloration; save the grey cement for black and dark grey tones. The best way to proportion the pigment is by weight: the pigment should never exceed 10% of the weight of the cement. Control the materials in the mix by weight. To assure uniform coloration, keep this weight proportion consistent from one batch to the next.

With this method you can pour the entire slab of colored concrete mixture or pour a conventional slab first, leaving the surface rough, then top it with a 1-inch slab of the colored mix.

Dust-on color. To apply color this way, purchase a ready-made, dust-on mixture or mix your own. Using the first product, prepare the concrete surface through the wood floating process; spread ⅔ of the amount specified by the manufacturer and float the surface. Apply the rest of the mixture, float again, and finish with a light troweling. Be sure to edge and regroove the joints after each application.

To mix your own, blend the pigment with grey or white cement (5 to 10 pounds of pigment to a sack of cement) in a dry state, then scatter the mixture over the damp concrete and finish as before. Another way is to mix pigment, cement, and sand in a 1:2:2 ratio. You will need about 50 pounds of this mixture to color 100 square feet (you'll use about 5 to 7 pounds of the pigment).

Brush-on color. This is the method used to apply color to the concrete surface after it has hardened. Apply the color in the form of a wax, stain, or paint. It makes no difference whether the surface is weeks or years old.

• *Waxes.* Specially developed concrete waxes containing pigments are easy to apply to a floated concrete surface (a hard finish, troweled surface is not porous enough). Clean the surface, brush or rub the wax on with a cloth, and allow it to harden. After the wax has hardened, rub it down with a hand or mechanical lamb's wool buffer. Areas that are subjected to a lot of foot traffic require rewaxing once or twice a year.

• *Stains.* The most durable of the brush-on colors, stains are inexpensive and are relatively easy to apply. For new concrete, allow it to cure at least six weeks. Clean the surface thoroughly, scrubbing with a wire brush and using tri-sodium phosphate (available at any paint store) in warm water to remove grease and stubborn soil. Flush with clear water.

Apply the stain with a brush or roller; add a second coat if darker tones are desired. Avoid painting the concrete when it has been heated by the sun. Semi-transparent wood stains work well on concrete.

• *Paints.* Paints provide the widest choice in colors. Water-base latex paint is one of the best kinds to use, but there are a number of others suitable for use on concrete. Clean the surface; etching with a 10 per cent muriatic acid solution never hurts. Brush or roll the paint on, applying two or three coats, depending on anticipated traffic.

How to Make Paving Blocks

CAST PAVING BLOCKS in ground mold: 1) dig 4-inch mold; 2) shovel concrete into mold up to rim below level of grass; 3) finish with trowel or wood float; 4) space for comfortable walking.

One of the best ways to make your paving an integral part of the overall landscaping is to use concrete paving blocks. You may know them as concrete stepping stones or concrete flagstones, but they are all cast concrete units, poured in small forms. The blocks may be poured right in location or made separately and placed later.

Pre-cast units are available commercially in a wide range of styles, colors, and sizes from most masonry building supply dealers. However, using the techniques for concrete paving described earlier, it is not difficult to make your own concrete paving blocks.

Casting in a ground mold

The simplest way to pour stepping stones is to pour them in a hollowed ground mold right where you want them. Dig out the desired contour for each stepping stone to a depth of 4 inches. Keep the distance between the excavations to 18 inches or less for comfortable walking. For economy, toss in some spare rocks to help make up some of the volume. Fill the excavation with a 1 part cement, 2 parts sand, and 3 parts gravel concrete mixture. Smooth the surface with a wood float—it is best not to trowel the surface as it will make it too slick, especially if it is set in a lawn. If stepping stones are placed in the lawn, their surfaces should be below the level of the grass to allow the lawn mower to pass over them.

Casting in a single mold

Individual concrete paving blocks can be cast in wood forms one at a time. This method is not recommended if you're going to pave an entire patio, but it is well suited for making stepping stones.

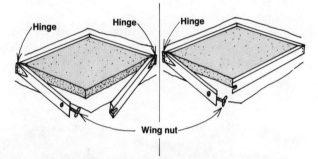

BUILD WOODEN FORM for casting single blocks. Pour concrete, let dry, remove form, and place block.

Build the form of 2 by 4s, cutting on angles at both ends so they fit together at right angles. Nail one corner, hinge two opposite corners, and fix the last corner so it can be secured with a bolt and wing nut. This type of form is the easiest to remove. Another (but less efficient) method is to nail two corners, hinge the third, and put a hook or wing nut on the fourth.

To pour the block, close the form and place it over a sheet of building paper on a square of plywood or mixing platform. After oiling the inside of the forms with a light motor oil, pour the concrete mixture into the form. Finish the surface with a wood float and let the concrete harden. Remove the form, clean it thoroughly, re-oil, and refill. The cast blocks should be kept continuously damp for three days.

Casting in a multiple grid form

Covering a large area with individually cast stones could postpone the completion of the project indefinitely. For a more expansive surface, consider using a grid of multiple forms and casting the stones in place.

Build a grid made of 2 by 4s in any angular pattern or in strictly rectangular shapes (the rectangular grid provides a more rigid form and is easier to work with). Finish the inner surface as smoothly as possible. Tapering the sides slightly toward the bottom with a plane will make it easier to remove the form. Don't make the grid so large that it will twist or wobble when you attempt to lift it from the hardening concrete.

To make and place the blocks, saturate the form with water and lightly oil the inside surfaces. Set the grid form down on graded and compacted soil and fill the compartments with concrete. Strike off the surface and finish with a wood float. When the concrete has hardened slightly, carefully lift the grid form off and edge around each paving block. Move the grid form to the next section in the paving scheme and repeat the process.

Curved paving blocks. For curved shapes, use a set of individual forms made of galvanized iron. Get a dozen or so strips 3 inches wide and bend them to shape; secure the ends with a cotter pin. Place the forms in the desired pattern and fill with concrete. Wait until the concrete has hardened slightly, then remove the forms, edge the blocks, and move the forms to the next section.

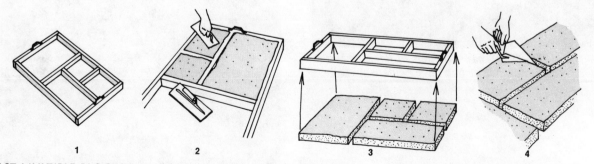

CAST MULTIPLE BLOCKS in grid form: 1) build wooden grid and place on desired location; 2) fill compartments with concrete and wood float surface; 3) remove form; 4) finish edges with 2 steel trowels.

ASPHALT PAVING

To many, the thought of asphalt as a patio floor material brings on visions of guests on a hot day slowly sinking, along with the patio furniture, into the sticky pavement like unfortunate captives of the La Brea Tar Pits.

Though it is still not recommended as the most ideal patio floor covering, today's asphalt is much improved over its ancestor which more closely resembled roofing tar. Asphalt paving can be put to good use, especially in paving driveways, garden paths, and service yards.

Paving with asphalt: pro and con

When you plan your garden paving, consider what asphalt will do for you and see if it fits in your plans.

Properly mixed, compacted, and installed over a solid interlocking pad of rocks, asphalt provides a surface almost as tough as brick or concrete. If not properly mixed and compacted, asphalt will likely soften in warm weather and show prints of furniture, cart wheels, or even shoes. Soft asphalt may stick to shoes and find its way onto the living room carpet.

Asphalt is very functional in driveways and service yards because the black surface does not show grease and oil stains the way other paving materials do. However, since the surface is so dark, it soaks up the sun's energy and radiates scorching heat. This characteristic makes asphalt an uncomfortable patio surface.

Asphalt paving is quite economical. Much less material is needed to cover an area than when you work with concrete. A 1½ or 2-inch topping

ASPHALT DRIVEWAY has shoulders of pebble-seeded concrete, gives asphalt strong edge, widens drive.

TROUGH of light-colored rocks carries rain water from asphalt driveway, gives good edge to paving.

on a good gravel or stone base will usually serve for driveways; 1 to 1½ inches for walks or patio floors. The surface may crack or sag if the soil shifts underneath it, but, unlike concrete, asphalt can be easily patched.

Kinds of asphalt

If you are considering paving with asphalt, you have your choice of two basic kinds: asphalt concrete and cold mix. The latter is more suitable for the home builder. Asphalt concrete is formed by coating heated and dried crushed rock with a hot asphalt cement. The asphalt is heated at the plant and is in a liquid state when it is put down. As it cools, it solidifies and binds the aggregates together. This type of asphalt is best left to the contractor who has the special equipment for compacting the rock fill and spraying on the hot mix. In addition, special equipment is required to transport the heated mix.

Cold mix. This type of asphalt is made by combining graded aggregates with either of two types of liquid asphalt—cut-back or emulsion mix.

The cold mix asphalts are well suited for patching. In most areas you can buy these types of asphalt in small amounts — usually in sacks of 60 to 100 pounds — for patching purposes.

• *Cut-back asphalt.* Volatile solvents hold the asphalt in solution in this mix. When the mix has been put down, the solvents evaporate, leaving the asphalt cement to hold the aggregates bonded together.

• *Emulsion mix.* An emulsifying agent in this mix holds the asphalt in suspension in water. The water in the emulsion evaporates, leaving the asphalt cement behind.

Emulsified asphalt. This asphalt is in a liquid state which is poured over a gravel base at the site. The asphalt emulsion sets up very quickly and is relatively easy to work with if its use is limited to a small area, such as a garden path, where you will have time to finish the surface before the emulsion sets.

The asphalt emulsion can be obtained in bulk from the asphalt storage lots located in many cities and rural areas. Don't try to work with asphalt emulsions in wet weather; a shower on freshly applied emulsion will ruin it.

WHITE DOLOMITE pressed into asphalt and on bed bordering driveway is helpful to driver at night.

ASPHALT PAVING in entryway provides subdued, non-glare surface, contrasts with white painted walls.

How to Pave With Asphalt

Asphalt paving must be uniformly compacted, usually with a heavy roller, and should be higher in the center allowing water to run off. For these reasons, asphalt is best applied in a continuous, uninterrupted surface, and it would be wise to limit the use of asphalt to smaller paving projects such as a path or small service yard.

SHOVEL COLD PLANT MIX in piles onto 4-inch base of compacted crushed stone or rock.

RAKE PILES of cold plant-mix evenly with garden rake to depth of 1 inch. Remove large rock while raking.

ROLL ASPHALT for smooth surface. Brush roller with water to prevent asphalt from sticking to roller.

Paving with cold plant-mix

You can handle this mix successfully if you limit its use to small areas where it can be hand-tamped or rolled. With smaller quantities of this mix, you will have to use a truck or trailer to haul it home from the mixing plant. Plants are geared to the volume needs of the paving contractors and may not deliver the smaller quantities you would need for the garden paving project.

Preparing the site. Outline the area to be paved with stakes and string. Remove grass and other plant growth; scrape down to firm soil and apply a soil sterilant. Because asphalt paving tends to crumble around the edges, install permanent headers around the area to be paved (see page 21 for the technique).

Laying the base. Put down a 4-inch layer of crushed stone or rock averaging ½ inch in size. Level and roll these with a heavy roller or compact with a power vibrator. (Both of these pieces of equipment can be rented in most areas.) Sprinkle additional rocks to fill in any remaining voids; compact. If you use cut-back mix, make the stone or gravel base tacky by priming with a light coat of liquid asphalt. For emulsion mixes, sprinkle the base lightly with water before applying the asphalt.

Applying the asphalt. Shovel emulsion or cut-back mix onto the base rock, raking it out evenly with a garden rake to a depth of 1 inch. Remove the larger rocks while raking. Don't rake too vigorously, for this will make the rock bed crumbly and unstable.

Finishing the surface. If you are using a cut-back mix, roll it continuously until it is smooth. When using an emulsion mix, roll it once, then wait about two hours and roll it again. To attain a smooth surface with the emulsion mix, scatter sand over the surface and roll it again.

For best results, the roller should be kept wet during the rolling process to prevent the mix from sticking to the surface and weakening the finish.

You can use a hand tamper to compact and finish the surface, but the results are not as good.

Limit the use of a hand tamper to very small areas.

This mix cures slowly, so treat it carefully for the first few months after installation.

What to buy. *For 100 square feet of 1-inch surfacing laid over a 4-inch rock or gravel base, you will need approximately these materials:*

Asphalt (emulsion or cut-back)	½ ton
Crushed rock (½ -inch)	2 tons
Sandy-binder or graded gravel	2 tons

Paving with liquid emulsified asphalt

Another way of laying asphalt is to put down a pad of graded rock, coat the surface with rapid-setting, emulsified asphalt, and cover the surface with sand. This is a fairly simple technique, well suited for home paving.

Prepare the rock pad the same way as described above for cold plant-mix. Apply the emulsion over the rock at a rate of 1½ gallons per square yard. Pour the mix from a sprinkling can or pour it from a square five-gallon can with the top cut out, tipping the can to produce an even flow. You can also rent a spray applicator; it's more expensive but will give the best results.

APPLY COATING of liquid asphalt to pad of graded rock; pouring from 5-gallon or sprinkling can.

Allow the first application to cure for 24 hours, then pour a lighter application of the emulsion mix (¼ gallon per square yard). Using a broom, brush on a coating of coarse sand or pea gravel (25 to 30 pounds per square yard); roll the surface thoroughly immediately following this application. Let the paving cure for another 24 hours. Then if the surface is not smooth enough

to suit you, repeat the emulsion and sand application. It is best to leave any surplus sand on the surface for a while to absorb any excess asphalt.

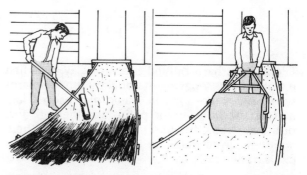

BRUSH coating of sand or pea gravel onto wet asphalt; roll thoroughly; let cure for 24 hours.

How much to buy. *To put down 100 square feet of paving on 4 inches of crushed stone with 3 applications of sand surfacing, you will require approximately the following supplies:*

Crushed stone	2 tons
Asphalt emulsion, rapid-setting	
Sand or pea gravel:	
one application	300 pounds
three applications	800 pounds

Coloring asphalt

Asphalt paving need not be black: it may be attractively colored with plastic paints specially manufactured for the purpose. Colors are soft in tone and range from light tan to dark green.

One product, frequently used on walkways and playgrounds, is a mineral-filled asphalt emulsion that gives a smooth, resilient, waterproof, non-skid surface. Manufacturers, however, recommend that the product be applied only by skilled contractors. You can obtain the topping in red, black, or green.

It is inadvisable to paint any asphalt surface that will be used for a driveway or a parking place. Oil stains will make it unsightly very soon. Gasoline and oil dissolve asphalt, so that you need a provision (such as an insert of gravel) to catch oil drippings. Without this, an asphalt surface used as a thoroughfare for cars may wear faster than a footpath.

One way to protect an asphalt surface from oil and gasoline drippings, as well as from weathering, is to coat it with a coal tar emulsion. You can apply it easily with a push broom; a gallon will cover about 100 square feet. The material comes in black only.

TILE PAVING

Tile, a thin, hard-burned piece of clay, was first introduced as a paving material by the Spaniards some 200 years ago. In California, padres laid hand-crafted tiles in the patios and the corridors of the missions, and the Dons who followed them adopted tile to decorate their comfortable, spacious haciendas.

Paving with tile: pro and con

For a formal and elegant look, tile is an excellent choice for garden paving. The warm, earthy browns and reds blend well with garden colors; its hand-fired pigments are permanent and non-fading. The tile itself is very durable. Smoother than almost any other garden surface, tiles are easy to clean and wax. Because of this, they are particularly well-suited to areas subject to soiling, such as an outdoor cooking area. Tiles provide a dressy surface that looks as well indoors as out—a good answer to a patio that extends into the house or a lanai. In most cases, tiles can be used to pave over a deck surface without affecting the supporting structures.

Tile paving does have some negative characteristics that should be considered. For example, tile is quite costly compared to brick—twice to three times as much per square foot. Because tiles are hard, they are difficult to cut and shape. The smooth surface of some kinds of tile can give off harsh reflections, especially when waxed. Tiles can also be very slick: around a pool they would give little traction to wet feet. To prolong the

SHADY, TILE PATIO made with foot-square 1-inch thick tiles laid on 2-inch sand base; edgings are wood.

QUARRY TILE set on mortar bed over wire mesh to pave low-level wood deck. Surface is easy to clean.

life of tiles, it is best to install them over concrete or wood decking, even though this involves additional time and expense.

Kinds of tile available

When you go shopping for tile, you'll have a wide choice of sizes, colors, and shapes. The tile you finally select will probably belong to one of two groups: quarry tile or patio tile—both manufactured products.

Patio tile is less costly than quarry tile and is more irregular in shape. It is also available in rough handcrafted shapes, reminiscent of the primitive tile put down by the early settlers. You can get patio tile in red brick and buff in these sizes: 12 by 12, 6 by 12, and 6 by 6-inch.

Quarry tile is more expensive and more regular in shape. The tile is available in tones that run from grey to brick color and in sizes 12 by 12, 9 by 9, 8 by 4, 6 by 6, and 4 by 4 inches. Outdoor tiles have a rougher surface than the glazed kinds normally used indoors and usually come in a ¾ or ⅞-inch thickness. They are available with both rounded edges and corrugated finishes.

TILE FLOORING extends from inside house out onto back porch. Extension of floor adds spaciousness.

FRIENDLY ENTRY is paved with large quarry tile set in mortar; ceramic tile was used in entry hall.

SMALL, INTIMATE PATIO is paved with large tiles. Brick-in-mortar edging separates patio from lawn.

CURVING BORDER around trees is made with rough textured double bricks. They look like tile but are easier to shape for curves.

How to Lay Tile

Laying tile is well within the capabilities of the home mason, but it cannot be a rush job. Because tiles are such a precise building unit, the slightest flaws in installation may be noticeable, so plan your work carefully.

Follow carefully the steps for grading the soil described in the beginning of this section on paving (see page 5); this is crucial for tile because the level must be accurate. If you're laying tile in mortar, take care not to stain the surface of the tiles with mortar.

Plan ahead

Take careful measurements of the area to be covered. Figure out how many tiles you will need to buy according to the size you have chosen (ordering a few extra tiles won't hurt). Be sure to allow for the mortar joints—¾ inch for large tiles, ½ inch for the small sizes.

If you find that you must cut some tiles to fill out a pattern or fill in an odd space, it is easiest to mark the tiles where they are to be cut and take them to a stoneyard to have them sawed with a diamond saw. If you have only a few to cut, you can chip them to size with a narrow cold chisel. Place the tile on a sack of sand, and, starting at the edge, nibble your way in to the line where you want it cut, using a series of sharp taps.

If your paving calls for a curving edge, cut the tiles in a series of angles that correspond roughly with the curve. You can also lay a curved border of brick on top of the rim to give the appearance of a curved edge.

Laying tiles in sand

For a less formal appearance (and only if you have stable, well-drained soil), you can lay the tiles on the ground or a bed of sand. If you aren't satisfied with the result, you can always set them in mortar later.

To lay tiles on sand, grade the soil and provide for drainage, leveling off the soil 1½ inches below the desired grade. Next, install edgings or headers, as described on page 21, around the perimeter of the area to be paved. Set the top surfaces of headers flush with the desired grade: slope for drainage 1 inch for every 8 feet. Make a screed by notching the ends of a long straight board—the depth of the notches should correspond to the thickness of the tile you will be using. The lower edge will then be the right depth for the surface of the sand bed.

Screeding. Pour sand between the headers and level it by moving the notched straight edge along the headers. If you make the sand bed more than ½ inch deep, the tiles may tilt when stepped on. One cubic foot of sand will cover about 20 square feet ½ inch deep.

Laying the tiles. You can lay the tile with open joints (½-inch spaces for small tiles, ¾-inch spaces for large tiles) or closed joints (tiles butted tightly together). Closed joints provide a more stable surface, since the tiles hold each other in place.

Lay the tiles in place starting at one corner. Give each a few taps with a rubber mallet to bed it in the sand. Check the level with a spirit level as you proceed.

A variation. For more stability, mix dry cement in a sand bed 2 inches deep. Level the sand as before, distribute dry cement evenly over the top (use 2 sacks of cement for each 100 square feet), mix the sand and cement with a rake, and level again, using the screed. For this method, the soil will have to be leveled 3 inches below the desired grade. Dust the surface of the sand-cement mixture with dry cement and lay the tiles dry as before with open joints. When the tiles are in place, wet the surface with a fine water spray. Don't let the water splash the dry mortar mix out of the open joints. Apply only enough water to make the sand-cement bed damp; avoid flooding it. Allow the surface to set for 24 hours. Then use jointing mortar to fill the joints (the procedure for this is described below).

Setting tiles in mortar

For a permanent surface, set your tiles in mortar.

Prepare the base as described above, allowing for a 1-inch mortar bed. Mix 5 parts sand with 1 part cement in a wheelbarrow or on a mixing board. Add water, continuing to mix, until you get a damp but stiff mortar mixture. Dump the mixture between the headers and level it, using the screed. Pour mortar in batches of 20 square feet or less, for you will have to set the tiles within an hour before the mortar hardens. Tiles should soak in water for 15 minutes before they are to be laid. Before setting them in place, stand them on end to let the surface water drain off.

Set the tiles in place with ½ or ¾-inch open joints, tapping each tile with a rubber mallet until it is firmly bedded in the mortar. Allow the tiles to set in the mortar for 24 hours before applying the jointing mortar.

Filling the joints. To fill the joints between the tiles, mix a 1 part cement, 3 parts sand mortar. Add enough water to make a soft mix that will almost flow. Pour this mix into the joints from a can with a spout or an old coffee can bent to make a pouring spout. Avoid pouring the soupy mortar mixture over the surface of the tiles; clean off any that does spill over immediately with a clean, wet rag.

Fill the joints flush with the surface of the tile or just below (no more than ⅛ of an inch). Using a piece of wooden dowel no less than 1¼ inches in diameter, compact and smooth the mortar joint by running the dowel over the mortar. Don't walk on this surface for at least 3 days.

Laying tile over concrete

To lay tiles over an existing concrete patio or walk, simply rough up the surface of the concrete by washing it with dilute muriatic acid solution. Clean and rinse the surface well.

Brush a coat of cement and water paste over the surface and apply the mortar bed as described earlier.

Laying tile over wooden flooring

You can use tile to cover stairs, a deck, or a porch with pleasing results. But first check to be sure that the wooden structure will support the added weight without sagging. If you have any doubt, install some additional supports before laying the tiles. Patio tile on a 1-inch mortar bed weighs about 20 pounds per square foot.

Nail a layer of waterproof building paper over the flooring. Stretch a reinforcing mesh of ¾-inch stucco or chicken wire and nail it down, leaving about a ¼-inch space between the wire and wood surface. This will help bind and reinforce the mortar bed.

Apply a 1-inch mortar bed using a 1 part cement to 5 parts sand-mortar mixture. Dust a layer of cement over the mortar bed and set the tiles in place with open joints as described above.

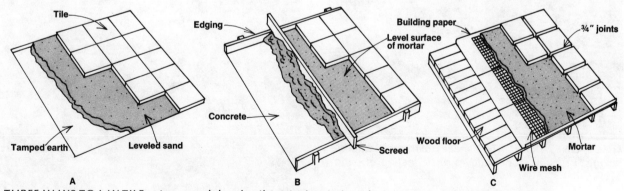

THREE WAYS TO LAY TILE: A) on sand, level soil 1½ inches below desired grade, install edging, lay tiles on ½-inch bed of leveled sand; B) in mortar, prepare base allowing for 1-inch mortar bed. Spread mortar mixture; level to 1-inch thickness; place tiles with open joints; fill joints with mortar; C) on wood, set tiles in 1-inch mortar bed over building paper and wire mesh.

FLAGSTONE PAVING

Pave with flagstone, and you're putting down the most durable surface available. Formed from compressed sand, clay, and other sediments, the rock from which flagstone is split has lasted and withstood the incredible wear and tear of massive earth movement and compaction through millions of years in geologic time.

Paving with flagstone: pro and con

Flagstone looks pleasant in almost any garden setting. It is one of the few paving materials which can be placed directly on stable soils. Its natural unfinished look blends well with garden plants. Not only do the subdued colors of the stone slabs—buff, yellow, brownish red, and grey—bring warmth to the patio but also their irregular shapes add texture to the garden floor.

Flagstone does have some less favorable attributes. It is the most costly paving material available, costing five to ten times as much as brick or concrete. Flagstones must be laid out very carefully to avoid ending up with a geometric mess looking like awkward patchwork.

Choices of flagstone

You have your choice of several different types of stone. Most commonly your choice will fall in the sandstone or slate category. Slabs are available in irregular or rectangular shapes. The latter can add even more to the costs because of the expense in cutting and shaping.

Flagstones generally range in thickness from ½ to 2 inches.

FLAGSTONES IN MORTAR makes extremely durable paving; irregular stones add to rustic feeling of patio.

INFORMAL GARDEN PATH of large flagstones set in soil; stones are spaced for comfortable walking.

How to Lay Flagstone

Flagstones can be laid on soil, on a bed of sand, or may be set in mortar over sand or in mortar over a concrete slab. Plan your work to minimize cutting and establish a pleasing pattern. As with most paving materials, provide your flagstone with a well-graded and drained foundation; follow the directions described in the introduction to this section on paving (see page 5).

Laying flagstone on soil

If your soil is stable and well-drained, you can lay flagstones directly on the soil. The stones you use for this should not be less than 2 inches thick in order to insure stability and strength. This paving is good for a walk but is not recommended for a large patio area.

Dig out the soil to a depth slightly less than the thickness of the stones and fit them into place.

Fill in the joints with soil and plant grass or set clumps of creeper plants.

Laying flagstone on sand

Flagstone on a bed of sand provides a more stable surface. The technique is about the same as that for laying brick on sand (see page 11) with a few differences.

Set up headers and put down and screed a 2-inch sand bed, laying the flags in place so that they are firmly bedded over their entire surface and won't wobble when walked on.

If you're using irregularly shaped stones, lay them all out first to get an idea of the overall pattern. Make any adjustments necessary before the final placement.

Fill in the joints with soil and plant grass or a creeper.

Laying flagstone on concrete

For the most permanent flagstone surface, set the stones in a bed of mortar on a 3-inch concrete slab. This can be an existing concrete patio. If you have to lay a slab of concrete, follow the directions given in the chapter on concrete paving (page 22). Allow the new concrete slab to cure for at least 24 hours before placing the flagstones.

First, lay out all the flagstones to try the pattern. Make any needed adjustments to get the desired effects.

Mix a 1 part cement and 3 parts sand batch of mortar (for technique, see page 68). Trowel enough mortar onto the slab for no more than 2 stones. Place the flagstones dry on the mortar bed and tamp them in place with the trowel handle or a rubber mallet and level.

Allow the stones to set for at least 24 hours, then fill in the joints with a mixture of 1 part cement, 3 parts sand, and ½ part fireclay mixed with water. Use a small pointing trowel to fill the joints and smooth the surface. Immediately wipe up any excess mortar spilled on the surface of the stones.

Paving with imitation flagstone

As a compromise to paving with the real thing, imitation flagstones can be cast in concrete with pleasing results. Color, pattern, and surface texture can be controlled to produce exactly the desired effect. Imitation flags are cast in a mold or a form. For the technique, see the chapter on concrete paving, page 34.

HOW TO CUT FLAGSTONE

If you're laying the flagstones in an area confined by headers or if you're having trouble establishing a pleasing pattern, you may find it necessary to cut and reshape some of the stones.

The easiest way to do this is to mark the stones, then take them to the supplier or stoneyard to have them cut with a special saw. If you have only a few to be cut, you can do it yourself using a brick set (broad chisel), a heavy soft-headed steel hammer, and a strip of wood. To cut the stone, mark the location with a pencil and, using the brick set, cut a groove along the pencil line with a series of sharp taps (wear safety goggles). When the groove is about ⅛ to ¼ inch deep, place the edge of the flagstone on the wood, put the brick set in the groove at about the middle of the stone, and split the stone with a sharp blow on the brick set.

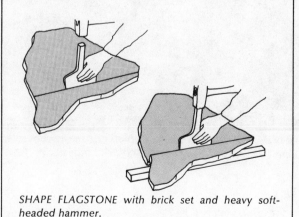

SHAPE FLAGSTONE with brick set and heavy soft-headed hammer.

STONES & PEBBLES

Pebbles and stones are the paving with a thousand faces. What screen immortal Lon Chaney did with Hollywood makeup, pebbles and stones do in garden surfacing. The effects you can create with natural pebbles and stones, by themselves or in combination with other paving materials, are virtually endless.

Pebbles and stones: pro and con

Larger pebbles and stones can be used to pave an informal garden walk or path. Properly placed, they look as much at home in the garden as they do in their natural surroundings—pebbly beaches, rocky fields, or boulder-strewn, wild-flower-covered meadows.

Stones and pebbles are widely available in countless shapes and sizes, are impervious to weather, and require virtually no upkeep. Set smaller stones and pebbles in concrete (or seed them in concrete; for technique see page 49). Large stones can be laid directly on soil as raised stepping stones. An entire surface can be paved

ATTRACTIVE WALK made with stones set in concrete; triangles were pre-cast, form colorful mosaic.

COBBLESTONE PATH made of large stones set in soil winds through charming, informal garden.

solid with cobblestones set in concrete, asphalt, or tamped earth. This paving is as practical now as it was in medieval Europe.

For small areas, try pebbles for making intricate patterns of pebble mosaic; use your imagination in designing a pattern. Narrow panels of pebble mosaic are very effective in breaking up an expansive concrete or brick patio surface.

There are some negatives to consider. Natural stones are very smooth and at times extremely slippery, especially in wet weather. Because their shapes are irregular, they are a slow and sometimes uncomfortable surface to walk on. This is especially true of the rounded cobblestones. Laying the surface, particularly when working with smaller pebbles and stones in mortar or concrete, is a very slow process. Try to confine this surfacing to a limited area.

When laying pebble mosaic, take care not to make the pattern too intricate; keep it simple.

GREY STONES set on soil in lush carpet of green moss form informal path in Japanese garden.

LARGE AND SMALL stones were pressed into concrete while still wet to achieve textured cobbled effect.

COLORFUL pebbles and stones are seeded in concrete; sizes altered from module to module to vary paving.

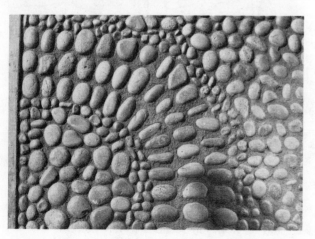

ATTRACTIVE MOSAIC of pebbles set in smooth mortar bed. Setting pebbles in sections varies pattern.

Setting Stones & Pebbles

The method used to set the pebbles or stones depends primarily on the size and shape of the stones and the effect you're trying to achieve. For larger (6 inches or more in diameter), flattened stones and a natural look, set the stones right on the soil. Smaller cobbles, river stones, or pebbles should be packed in a mud bed or set in mortar or concrete. The latter method of setting stones requires installation of headers or other forms of edging (see page 21).

If you want to make a mosaic, plan it carefully. This is especially important if you have to pave a large area. It is always a good idea to draw out a preliminary sketch to get a better idea of what the final design will look like. In most cases you can draw your design to size on wrapping paper; for large sections you may have to do it in sections. Try to group stones by size and color whenever possible. If you want to mix sizes and colors, make certain to mix them as randomly as possible to avoid awkward designs.

Laying stones on soil

Lay large stepping stones directly on soil or bed them in soil to any desired depth—even flush with the ground.

Stones laid on the ground should be fairly flat so they will not wobble or rock when stepped on. Depending on their size and desired visual effect, space the stones to accommodate a comfortable walking pace. When the stones are in place, check to make sure that they are secure and stable. If they wobble, pack soil, gravel or small stones underneath to stabilize them. It helps to loosen and soak the earth before placing the stones and then to press the stones so the ground will conform to the bottom surface of the stone.

To embed stones in the ground, dig to the desired depth, put down a ½-inch bed of sand or gravel, and place the stones as before.

Setting stones in mud

Setting in mud is the procedure that has been used in Spain for centuries. Some people prefer this traditional setting to cement, mortar, or concrete because the mud base harmonizes better with the pebbles or stones. If the clay is properly prepared, it forms almost as permanent a base as mortar.

Set up your edgings (see page 21), if needed,

and prepare the base with a 2-inch gravel pad topped with 1 inch of sand.

Screen dry clay soil until it is as fine as dust. With the dust, fill in the area where the pebbles or stones are to be inserted.

Using a fine mist of water, wet the clay a small section at a time until it is the consistency of dough (a hard spray will send the clay dust billowing in all directions).

Set the stones close together and on end in the mud, a little high at first, then press them level with a board to embed them evenly and firmly. Loosen up the clay soil and water soak the surface. This will drive out all the air. When dry, the clay will be almost impervious to water.

Setting stones in mortar

For permanence, set stones in mortar, on a slab of concrete, or in concrete itself.

To set the stones in mortar over concrete, pour a concrete slab as described in the chapter on concrete paving. (The concrete surface can also be an existing patio.) If you're pouring a fresh slab of concrete, allow it to set for at least 24 hours before setting the stones on top.

Spread a 1 part cement and 3 parts sand mortar mix ½ inch deep over the concrete slab and level it. Stones should be set in the mortar within two hours, before the mortar sets. If the area to be paved is too big to follow this time table, set the stones in sections. When you mortar the stones in sections, spread a mortar bed you can

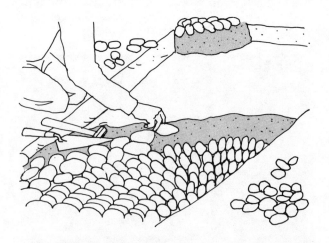

SET STONES in mortar over concrete slab. Spread mortar and set stones in sections to vary pattern.

fill within the 2 hours, then cut the dry edges away from around the previous mortar bed before you spread the next section of mortar. Mortaring in sections allows you to vary your pattern with each section.

Keep the stones in a pail of clean water, setting them in the mortar when they are still wet. This prevents the stones from sucking moisture out of the mortar, a process that would substantially weaken the mortar.

When pressing the stones into the mortar bed, push them down deep enough so that the mortar gets a good "hold" on the edge of the stone—generally just past the middle where the circumference of the stone begins to narrow. Use a board to keep the stones level. When the mortar has set (after 2 to 3 hours), spread a thin layer of mortar mix over the surface and into the voids, then hose and brush excess mortar away before it sets.

USE STRAIGHTEDGE placed across top of stone paving to keep surface level as stones are set in mortar.

Setting stones in concrete

Stones or pebbles can be set in concrete two ways: 1) seeded, pressed down beneath the surface of the wet concrete, and later exposed; or 2) usually in the case of larger stones, set in wet concrete one by one and left exposed.

Seeding. This method is generally used when you're working with smaller pebbles. Set up forms as described on page 21 and in the chapter dealing with concrete paving (page 29); pour the concrete and level the surface. Now spread the pebbles over the wet concrete and press them into the concrete with a wood float. Pebbles should be completely covered. Take care not to

over float (see page 32 for floating technique) the surface, since this buries the pebbles too deeply in the concrete, making later exposure difficult. When the concrete has set to a point where it will support your weight on a knee board without indentation, wash the surface with a fine spray and scrub with a firm brush to expose the pebbles.

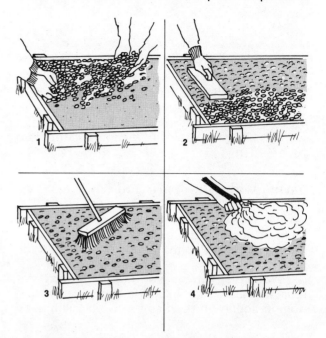

TO SEED CONCRETE, 1) spread pebbles evenly over wet concrete; 2) push pebbles into concrete with float; 3) and 4) alternate brushing and hosing to expose seeded pebbles in concrete surface.

Handsetting stones. For larger stones, pour the concrete surface as before and push the stones one by one into the concrete, covering slightly more than half the stone. Don't fill your forms completely with concrete because it will overflow the forms once you begin pressing the stones into the mix. When the concrete has hardened somewhat, the stones can be exposed to any degree by hosing and brushing.

A variation. If you prefer to have the stones set absolutely level, set them upside down in a 2½-inch-deep form with a plywood bottom. Place the stones butted together, best side down. Pour a layer of dry sand ⅜ inch thick between the stones; fill the rest of the form with concrete. When the concrete has set, turn the mold over, remove the wood form, and brush out the sand.

This same method is used on a much grander scale for walls of buildings set with pebbles. Entire wall sections are set in a huge form as described above. When the concrete has set, the entire panel is raised into place, braced, and secured.

WOOD SURFACING

To bring nature and informality together in the garden, pave your outdoor living area with wood. Few materials can match the natural rustic qualities of wood—its warm color and soft texture seem to bring something of the forest carpet into your garden.

Paving with wood, pro and con

Paving with wood has many advantages. Wood looks at home in the garden; a wood surface is not apt to clash with the garden landscape or the plants.

Wood is an extremely workable surfacing material that allows the builder a good deal of freedom in application and construction techniques.

Soil or drainage problems that would make concrete or brick surface difficult to install can be overcome with greater ease when paving with wood simply by elevating the surface above the ground in the form of a low level deck. Coloring your deck to match the house or the landscape is relatively simple with the many long-lasting wood stains now available.

Wood can be very pleasing and effective in combination with other surfacing materials. Break up an expansive surface of concrete or brick by using wide edgings or headers made of heavy timbers or railroad ties. Add variety to a loose gravel surface with a pattern of redwood rounds.

Paving with wood does have some drawbacks. The most critical of these is the impermanence of

REDWOOD ROUNDS laid in gravel lead through garden gate to front door; setting has Japanese touch.

ROWS OF WOOD rounds laid on soil in bed of smooth round stones; walk has friendly, woodsy feeling.

wood. In climates where rain, snow, and freezing conditions are minimal, wood surfacing will probably last for at least 5 and maybe as much as 10 years. But frequent rains, snow, and freezing conditions can limit the service of wood surfacing to as little as two to three years. Because decking is raised from the ground and dries out quickly, it will last longer under these conditions. Wood rounds and blocks are usually set in place so that the end grain is in contact with the ground moisture that seeps up through the soil or sand bed. The open grain soaks up the moisture like a sponge. Moisture provides an excellent environment for bacterial and insect growth; the wood eventually either rots away or succumbs to insect invasion.

Choices of wood paving

When you pave with wood, you can choose from a variety of products, and you can vary the way the product is used.

One of the most natural looking wood products is the wood round. The disc-like rounds are cross-section slices of a tree trunk—usually redwood, cedar, or cypress trees—and vary in diameter from 6 inches to 4 feet.

You can also purchase wood blocks, usually sliced from railroad ties or large timbers. These are quite hard and can be laid like bricks.

Board lumber is the most versatile wood product for paving. Use it to build a low level deck or lay short strips of lumber in parqueted fashion. An interesting variation is to build a low level deck out of modular units. These are small sections of decking laid together with the option of being able to rearrange them later.

How to pave with wood

Paving with wood is not difficult but does require careful planning and advance preparations. For blocks and rounds placed on the ground, you will have to prepare the base to allow for good drainage. Above-ground paving such as decking requires solid footing—supports placed in or on firm ground that will not give under the weight of the structures.

Treating the wood with a preservative is another essential preliminary step, adding to the life of your paving. A number of excellent solutions are available. Your lumber supply dealer can recommend the type best suited to your needs. Using a preservative that contains a toxic agent will discourage insect invasion.

(Continued on page 54)

GARDEN PATH is paved with wooden blocks set in mortar. Color and texture blend with forestlike setting.

STRIP OF WOODEN BLOCKS breaks up expansive concrete driveway; set blocks like bricks in sand or mortar.

STEPPED WALKWAY is paved with varying sizes of redwood rounds; they range from 6 to 36 inches.

RAILROAD TIES, 8 feet long, are laid flush on a 2-inch sand base to form both walk and steps. Gaps between ties were filled with fine sand. Lay ties lengthwise for narrower walks; stack them for walls.

SMALL, LOW-LEVEL DECK is extension off living room; narrow section of decking leads to concrete patio.

WALKS PAVED with railroad ties; ground was dug away so ties sit level with ground; mortar fills joints.

NARROW LOW LEVEL deck extends width of yard, provides additional outdoor living space off concrete patio.

DECK MODULES form ground level surfacing. Modules are easily placed and rearranged to vary surface.

WOODEN DECK of 2 by 4s is set flush with surrounding lawn and concrete patio, makes good playing court.

ENTRY WALK of 2 by 4s nailed to 2 by 4 runners on sand looks like old-time western sidewalk.

End-grain blocks

Wood blocks can be laid in much the same manner as bricks. Clear the area to be paved and excavate to a depth 1 inch greater than the thickness of the blocks. Set 1 by 4-inch headers around the circumference of the area. Tamp the ground and cover with one inch of sand. For extra protection from ground moisture, before adding the sand, put down a sheet of polyethylene plastic with holes punched one every square foot for drainage.

Place the blocks level with the edging with 1-inch open joints; adjust the joints if necessary to fill out the desired space. When blocks are in place, tamp the sand down into the joints with a narrow piece of wood.

Another method of laying blocks is to set them in a 1 part cement, 3 parts sand-mortar mixture. Follow the procedures outlined for brick paving, page 14.

Paving with wood rounds

Rounds are easily set in place. Clear the area and excavate and grade the soil as before. With rounds, too, it is advisable to shield the paving from extensive moisture seeping up from the ground by covering the area to be paved with plastic sheeting before putting down the sand bed. Puncture the plastic for drainage.

Put down a 2-inch sand bed and place the rounds on top in random style. Alter your pattern of large and small discs to fill out the desired area and to attain a pleasing effect.

Fill the spaces between the rounds with soil that will support plant growth or fill with loose aggregates such as wood chips, gravel, crushed brick, or pebbles.

For a more permanent flooring, set the rounds in mortar. Keep in mind, however, that if the rounds wear away or decompose, the entire surface will have to be torn out.

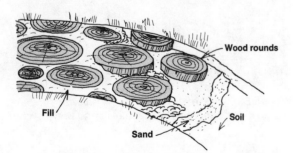

PLACE WOOD ROUNDS in a bed of sand over graded soil. Fill spaces with soil for plant growth.

Paving with railroad ties

Railroad ties can be used very effectively as a surface in a limited area. To pave an entire patio with ties could prove rather costly (used ties, 6 by 8 inches by 8 feet, can cost up to $5 per tie). Railroad ties look very good with other paving materials. To pave with ties, clear plant growth from the site and sterilize the soil. Lay the ties on top of the soil or on a bed of sand over sheets of polyethylene plastic. Another way is to dig down so that the ties may be laid flush with the ground.

Laying a parqueted floor

Where drainage might become a problem, a raised parqueted floor could be the answer. Build a support grid out of pressure preservative treated 2 by 4s laid on tamped earth or over plastic and tamped sand at 28-inch intervals in both directions. Using galvanized nails, nail 7 31-inch 2 by 4s onto the grid, as shown below, overlapping the grid 1½ inches on both ends and spaced to cover 31 inches. The next section of 2 by 4s is nailed onto the grid in the opposite direction.

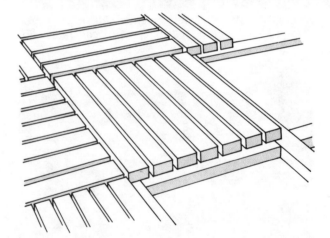

BUILD PARQUETED FLOOR of short 2 by 4s nailed to supporting grid of 2 by 4s set 28 inches apart.

If the surface of the flooring is below the level of the surrounding garden, lay drainage tiles around the perimeter in a gravel-filled ditch about 6 inches below the ground.

Paving with modular decking

The principle of parqueted wood paving can easily be applied to the construction of modular decks. Instead of covering a 2 by 4 grid which has been laid on the ground, cover individual 2 by 4 square frames to form separate movable modules. Place these on graded soil or a bed of sand in any pattern desired. Omit a square here or there to accommodate a container plant. Later, when you plan an outdoor party, move the modules around to provide a large uninterrupted surface. An added feature of this kind of paving: the modules can be constructed indoors and used when and where desired. Use all the modules for a large surface or take up a number of them and store them. They fit easily into a limited storage space.

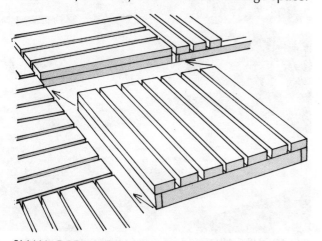

SMALL DECK MODULES can be any size. Place on graded soil or bed of sand in any desired pattern.

Paving with low-level decks

A low-level deck will give you a solid, relatively durable garden surface requiring little or no earth moving or grading and a minimum of maintenance once it is installed. One of the best ways to extend an indoor room, a low-level deck can be built extending out into the garden without a change of level.

Set pre-cast concrete supports on tamped soil that has been treated with a soil sterilant to eliminate undergrowth or set them in the ground to desired depth. Place the rows of blocks no more than 3½ feet apart; the interval between blocks in a row should be about 4½ to 5 feet.

Lay 4 by 4-inch pressure preservative treated support beams across the top of the concrete foundation blocks. Level these by inserting shingles or other wood scraps between the beams and supports. Toenail the beams to the blocks. Joints between beams should meet and rest on the concrete supports. Secure the joints by bolting or nailing metal plates or wooden braces across the sides of the joints.

Lightly nail a 2 by 4 across the end of each beam to keep them straight and in place. Then lay and nail in place the rest of the 2 by 4 deck

boards. Use galvanized 8 penny nails. Space the boards 3/16 inches apart to allow for drainage (a short length of wooden yard stick with a nail driven through it will act as a good spacer and assure consistency throughout).

Low-level decks such as the one described here can be built by the home builder without too much difficulty. For more complex structures such as those extending from the house out over a sloping hillside, you will need detailed plans and perhaps some professional advice. (See the *Sunset* book *How to Build Decks.*)

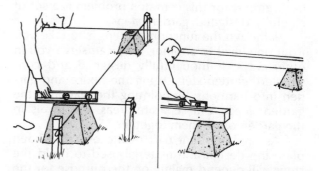

SET PRE-CAST CONCRETE SUPPORTS on tamped soil or in soil to desired depth; check level. Lay 4 by 4 support beams across footing and check level.

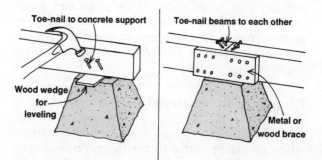

TOE-NAIL SUPPORT BEAMS to concrete footing. Use small wood wedges to level base. Toe-nail and brace joints between beams over concrete support.

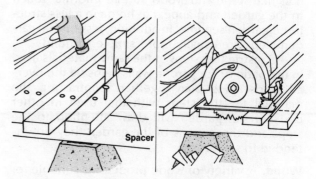

NAIL 2 by 4s across 4 by 4 support beams; use 3/16th-inch spacer to insure even spacing. Trim ends of 2 by 4s to even length with circular saw.

GARDEN STEPS

You are bound to have your ups and downs when landscaping or building in the garden, and the best remedy for this common problem is a set of carefully designed garden steps.

Although the function of steps is to take you from one level to another, they can serve you in other ways. If thoughtfully designed and constructed, garden steps bring the house and garden into harmony by pointing the way into the garden. Steps can separate areas and levels in the garden or tie them together.

As an important part of the overall garden plan, the design and material used to build the steps will depend mainly on the purpose for the steps. Narrow, steep steps might be appropriate on a rarely used bank or combined with a wall for privacy on terrace. Wide, deep steps are more inviting and make an excellent transition from a parking area to an entry. Wide steps can focus on the focal point or living area of the garden, give a sense of spaciousness, or serve as a retaining wall, a base for a planter, or additional seating space.

Choosing the material

Consider the house and the rest of the garden when choosing materials for steps. Poured concrete and masonry block units such as clay bricks and concrete blocks usually present a substantial and structured look. They are generally more at home in the formal garden setting. Natural materials like stone and wood add an informal touch to the garden and appear at home in a less structured garden scheme. In addition, less formal garden steps are usually more simple to build.

Constructing steps of the same materials used in the patio or for garden walls helps tie the overall landscaping plan together. Contrasting the material for the steps with that used in other parts of the garden will draw attention to the steps and those areas of the garden they are intended to serve.

Wood. A variety of wood products is suitable for step building. Although woods such as redwood and cedar are naturally resistant to rot and insect invasion, plan to use an additional preservative

GENEROUS GARDEN STEPS used railroad ties for risers that also retain gravel-covered soil treads.

on these woods, as well as on all wood that comes in contact with the soil. Pressure-treated wood products offer additional protection and durability. Wood can be used both for the tread (the horizontal surface of the step that you walk on) and the risers (the vertical surface that rises to the next level). Wood can be used effectively in combination with masonry, gravel, or ground cover tread. Various patterns of redwood, cedar logs, or railroad ties give a rustic effect.

Railroad ties, logs, and rounds can be set directly on the soil but require cutting and filling to provide a stable surface. Ties and logs can be used across by themselves or to confine masonry, gravel, or soil to support ground cover treads. Ties can also be stacked lengthwise side by side. Use your imagination to create your own combinations.

Stone. Complementing a natural setting, stones can be used in a variety of shapes and sizes. Use small, smooth stones set on mortar or concrete for a cobbled effect. Or place single large, flat stepping stones stacked to support one another. Field stones or broken concrete pieces can be laid dry or set in mortar. Keep in mind that treads or stones set in mortar require sturdy forms.

Bricks. Use bricks set in mortar confined by wood headers or stack the brick to form both the risers and the treads. You will have to cut the slope and provide the bricks with a sturdy foundation. Bricks may also be laid on sand confined by wood headers.

Concrete. All forms of concrete are suitable for building garden steps. Concrete can be cast in place, using a series of staggered forms. Precast concrete slabs of various sizes make an attractive, contemporary change in level. Concrete blocks may be laid on sand confined by headers or in mortar in much the same way as bricks.

Basic considerations

For building garden steps, follow the same construction practices as you would for building walks, patio floors, or garden walls. The tread of the steps must be provided with the same kind of foundation and support you would give to paving. The risers frequently function as mini-retaining walls holding in place soil or the materials used for the treads.

Design. When designing your steps, don't assume that they must cover the shortest distance between two levels in the garden. Steps, like walks, can afford leisurely enjoyment of the garden by leading you up, down, and around the landscape. Sizes of the treads and risers and their relationship should be based to some extent on who will be using them.

Riser and tread relationship. The relationship between tread and riser is an essential consideration for beauty and function, as well as for safety and comfort. Generally, the closer the treads and risers are spaced and elevated to accommodate a normal walking step, the better.

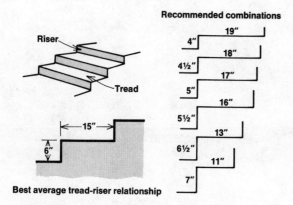

PLAN TREAD-RISER RELATIONSHIP based on beauty, function, safety, and comfort. These are best.

Landscape architects and engineers recommend a 6-inch riser and a 15-inch tread as a comfortable average. As a general rule, twice the riser plus the tread should equal 25 to 27 inches. A ramplike stairway might use a 4-inch riser with a 28-inch tread, while the steepest combination for the garden stair would be a 7-inch riser with a bare minimum 11-inch tread. Indoor dimensions of stairways do not correspond to the larger scale of outdoor dimensions.

Length. The change in level of your slope and the materials to be used in the steps will be prime considerations in the design of the steps. You can measure the level change by setting up a measuring device like the one shown above. The distance from A to B is the change in level; the measurement from A to C is the minimum distance your steps should travel.

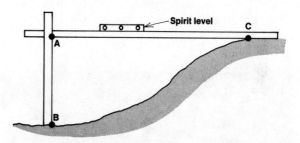

SIMPLE DEVICE measures change in level of slope.

Building the steps to fit the slope usually results in a bad tread-riser relationship. You will probably have to cut the bank or fill under the steps to make the slope fit the stairway.

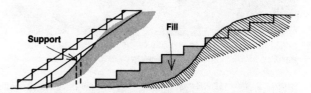

CUT BANK and fill under steps to fit slope to steps.

Width. Although the width of the steps should fit in with the rest of the garden, a minimum width should be 4 feet to accommodate one person and 5 feet to accommodate two persons. Purely functional access steps could be as narrow as 2 feet, but, before building, consider what kinds of equipment you might want to move up the steps. A ramp might be desirable next to the steps for wheelbarrows and lawnmowers.

If you want a random edge on your steps, don't include the staggered section in the usable width. No tread should be less than 11 inches wide.

First, a plan

Work out the plan on paper drawn to scale, keeping in mind that an unbroken steep slope is seldom a good idea. The distance can be broken with landings or by curving or zig zagging the steps. If you have doubts about your design, you might like to rough them out on a 1 by 12-inch board. Mark the steps with stakes in the ground, adjusting them until they are even.

Construction techniques

Steps using graded soil for treads are probably the easiest to build. Use wood, brick, or concrete for risers. These should be held in place with stakes, pipes, or pins through drilled holes. Gravel or ground covers, such as dichondra or are-

naria, make good treads with this type of step. If you want to use grass, make treads wide enough for cutting ease. This type of step could be a temporary one; later, the tread could be built of wood or masonry.

Basic wooden steps can be constructed without stringers by using stakes that are nailed to risers, then positioned and driven into the ground. Timbers such as railroad ties can form both the tread and the riser, leaving little else to build. Be sure to set them firmly in place so they don't wobble.

Support is a prime consideration for masonry, as well as for wooden steps. Use reinforced concrete when there is much fill underneath. If you don't use reinforced concrete, the subgrade should be firmly compacted by tamping.

For a rough surface on concrete, expose the aggregate in the concrete mix by hosing and

NATURAL LOGS form risers of this gently-climbing garden walk; treads are pea gravel over soil.

SPLIT, ROUGH LOGS are pegged into slope and edged with cedar rounds to form rustic garden steps.

BRICK RISERS set in concrete retain gravel treads, ease walking up gentle slope.

CONCRETE BLOCKS stacked in overlapping fashion form treads and risers; block steps are easy to curve.

scrubbing the treads and risers before the concrete hardens. This should be done immediately after the forms are carefully removed; you will not have much time, particularly if the day is warm. If you are working near the house, be sure to protect your foundation with plastic sheeting before hosing.

With such smooth surfaces as quarry tile treads, you may want to pitch the tile to the back ¼ inch for wet weather safety.

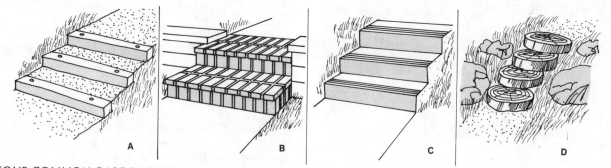

FOUR COMMON GARDEN STEPS: A) headers placed down the slope; soil or gravel fill forms treads; B) masonry blocks supported by wood, steel or masonry supports laid in mortar on cut slope; C) steps of concrete cast in place; poured in wooden forms; D) wooden blocks, flat stones, or concrete stepping stones; stack in overlapping fashion down the slope.

RAILROAD TIES set into bank and backed by gravel make rustic wide steps between garden and woods.

USED BRICKS set in mortar form risers and treads of broad steps leading from rustic terrace to deck.

STEPS of reclaimed material: treads are used cobbles set in mortar, riser frames are railroad ties.

CONCRETE BLOCKS mortared together form permanent garden steps. Use building blocks or paving blocks.

STAGGERED STEPS of poured concrete have exposed aggregate finish for permanent, non-slip surface.

BROAD FLIGHT of exposed aggregate concrete steps welcomes you from parking area, points way to house.

WOODEN RISERS and concrete treads form casual steps leading up to terraced organic vegetable garden.

ATTRACTIVE CONCRETE steps are seeded with colorful, smooth pebbles; wooden forms were left in place.

SHALLOW CONCRETE steps with exposed aggregate finish blend into lush bank of greenery.

STACKED STONE SLABS curve down steep bank to lower terrace from concrete patio above.

PLANK STEPS with ends staggered wind up to entry-way. Both risers and treads are 2 by 6s.

NATURAL STONE STEPS are almost hidden by creeper plants. Plant growth in cracks tie steps to landscape.

WOOD UTILITY POLES can be cut in sections, set in the ground to form attractive, informal garden steps.

LONG SLOPE made navigable with circles of pre-cast, pebble surfaced concrete set on recessed pedestals.

STACKED INTO SLOPE, railroad ties make bold steps connecting walk with patio of stones and concrete.

Walls

The value of the free-standing wall was recognized ages ago when builders spent much time and effort erecting such walls primarily for protection around estates, entire towns, and even countries. Around 200 B.C., Chinese builders erected a monumental defensive wall that stretched 1,500 miles between the Chinese empire and Mongolia. This wall still stands today as a testimony to the strength and durability of masonry wall construction.

Though heads of state may never visit your garden or patio wall, it has much in common with the Great Wall of China: it serves as a boundary, encloses an area, and provides privacy. But these are just three of the most obvious functions; there are many more.

Low walls are used for auxiliary garden furniture, to separate sections of the garden either decoratively or functionally, or tie together related elements of a garden plan. A wall can be the focal point of the garden or may be used to define the edge of a flower bed. Combine a wall and water as a part of a reflecting pool or waterfall. Use one as part of a barbecue or outdoor fireplace.

The higher masonry wall, traditionally built to mark the boundary of an imposing estate, can work effectively to deaden the noise from a busy street, enclose an outdoor room, or protect from the merciless rays of the sun in hot climates. The wall may be solid or open, constructed from one of a great variety of decorative concrete screen blocks. A masonry screen allows light into the patio area but provides welcome shade in hot summers and allows privacy without entirely blocking out the view.

The inherent strength of a wall makes it an excellent soil retainer. Many homes being built today are set on hillside or sloping lots that present a unique landscaping problem: how to make the slope useful and keep the earth in place during heavy rains. The answer is the retaining wall. Whether it is only a couple of yards of soil in a raised bed or an entire hillside that must be kept at bay, a carefully constructed wall will do the job.

Choose from a variety of materials

Building supply yards today stock a great variety of materials suitable for use in wall construction. Your choice among these materials depends largely on the function and location of the wall and how much time and effort you plan to invest in its construction.

Masonry blocks. For a solid wall and clean structural lines with pleasantly textured facing, you can't beat block walls built of bricks, concrete blocks, or adobe blocks. For the amateur builder, these materials are relatively easy to handle. But they do require mastering the art of building with mortar.

If you build with brick, you can have a choice of many patterns. They require less effort to handle, but because they are relatively small, brick walls will take longer to erect.

Concrete blocks can also be laid up in a variety of patterns. And with this material you get a tremendous choice of shapes and sizes. Decorative screen blocks, available in many designs, allow the builder to vary the design in the wall. Because concrete blocks are bigger than bricks, a wall built of them will take less time to build.

A wall of adobe blocks brings a bit of the old Southwest into the garden. They are laid up in much the same way as bricks but are about five times as big and require much more effort to handle. Adobe blocks are not as readily available as clay bricks or concrete blocks.

CURVING WALL separates side garden and entry patio from the street. Wall serves as boundary, provides privacy.

Block construction makes planning simpler and more exact.

Natural stone. Constructing a wall of natural stone offers a real creative challenge to the builder. The finished product adds a natural look to the garden in a way that few other building materials can. Supply yards usually stock a variety of stones, shaped and unshaped. If your home site contains a supply of natural stones, you can put them to good use in a stone wall.

Stones can be laid up dry or with mortar. But because of their inherent uneven shapes and sizes, they are considerably more difficult to build with than the manufactured blocks described above.

Poured concrete. Pouring a wall of concrete permits a great deal of flexibility in the design of the wall. Plastic concrete is relatively easy to place—up to a certain point. Above 3 or 4 feet, it calls for experienced hands and a lot of heavy equipment. Much of the work involved in pouring a concrete wall occurs in the preparatory stages, primarily in constructing the forms.

A poured concrete wall is among the strongest you can build and is especially helpful in retaining soil.

Wood. Walls of wood are used in the garden primarily for retaining purposes. Wood is an excellent choice for building the low walls of raised flower beds.

For purposes of wall building, you have many wood products suitable to choose from. Board lumber is easy to work with, railroad ties look bold and rustic, natural logs bring a bit of the forest into the garden, and telephone or utility poles set upright in the ground add a vertical line to the horizontal landscape.

Construction basics

Include a trip to your City Hall in your planning itinerary to find out if a special permit is required for the wall you wish to build. Be sure to check your local building codes which give height, setback, and other structural requirements. If the wall you plan to build will eventually be used as a load-bearing wall, you will need to secure permission from the city. Generally, however, if your wall is going to be less than 6 feet in height, special permit is not required.

Foundation. Regardless of the type of wall you plan to raise, you will have to provide it with a solid foundation (unless you're building with wood).

Poured concrete is about the best foundation you can provide for a wall because it can be smoothed and leveled better than most other materials. The procedure for installing a concrete foundation is the same as that for pouring a concrete walk or patio floor (see page 22). For most walls, a 12-inch-thick foundation is best. Concrete may be poured into wooden forms; their straight wooden edges make it easier to level the plastic concrete.

From an appearance standpoint, it is usually best to keep the surface of the foundation just below the surface of the ground. This will give the appearance of the wall rising out of the ground rather than sitting on top of it.

For very low walls no more than a foot in height or for low raised beds, you can lay the base of the wall directly on the soil or in a leveled trench.

BRICK WALLS

Much of the quaintness of European architecture comes from the use of brick. Handcrafted brick walls with neatly tooled mortar joints give even the simplest country cottage a warm, distinctive look. Builders have erected walls for thousands of years, and bricks are as popular in wall construction today as they were when first used almost 5000 years ago.

Building with brick: pro and con

For the beginning mason who has never built a wall, bricks are an excellent material to start with. They are easy to handle and place, and their uniform size makes planning the job simpler.

Bricks are available in many colors and textures; you'll find a brick to complement almost any garden plan. Brick ages naturally, changing color gradually over a period of years and acquiring a rustic patina.

Lay bricks in the form of a solid wall or stagger them, leaving open spaces to create a friendly screen effect. Because bricks are a small building unit, they lend themselves well to curving construction. Slightly angling each brick produces a smooth, serpentine structure.

Bricks do have some less favorable characteristics. Because of their small size, it will take considerably longer to erect a brick wall than if you were using larger concrete blocks. Though professional bricklayers can lay up to 1,000 bricks a day, the amateur will normally have to settle for about a third of that.

A single thickness of brick does not have the strength of a single row of concrete blocks or a 4-inch-thick (the width of one brick) poured concrete structure. A brick wall of this thickness will have to be limited to about 2 feet in height.

Choose from a variety

Brick manufacturers today provide the builder with a large variety of colors, textures, and

LOW BRICK WALL retains soil in raised beds; plantings at various levels add vertical dimension to garden.

RAISED PLANTING bed built with used brick; low wall appears to pass through massive natural rock.

RUSTIC USED BRICK wall separates garden and patio area from swimming pool; pilasters reinforce wall.

strengths in bricks from which to choose. Basically, bricks fall into two categories: 1) common brick, and 2) more expensive face brick. Common bricks are less consistent in size, color, and texture than face bricks. Face bricks are made of especially selected materials and fired harder than the common brick, making them stronger. The hotter the temperature at which the brick is fired, the greater the resistance to adverse weather conditions. Face bricks are used more often in wall work.

Weather considerations. National specifications classify bricks in three grades: Severe Weathering (SW); Moderate Weathering (MW); and No Weathering (NW). For areas with severe climate variations, especially where bricks will be exposed to heavy rain and frost and come in contact with the ground (as in retaining walls), the builder should use SW grade. Choose a MW grade of brick where moderate resistance is required; these are suitable for free-standing walls. NW grade bricks should be used only as back-up for interior masonry where it will not be subjected to freezing and thawing cycles. Consult your building supply dealer for the proper grade in your area.

Many people prefer to build with used bricks. These are taken from old buildings and walls and are frequently marked, chipped, and stained with mortar. Used bricks can add a pleasant rustic look to a garden wall. Because their supply is limited in some areas, a number of manufacturers convert new bricks to used ones.

(For more detailed descriptions of brick varieties, see the section on brick paving.)

How to buy bricks

Visit your building supply dealer and decide on the kind of brick you wish to use. Check a few other suppliers, as well, because prices can vary somewhat from one supplier to the next. The number of bricks and amount of mortar ingredients you buy will depend upon the pattern you use for the face of the wall and the thickness, height, and length of the wall.

When you arrange for delivery of the bricks, consider paying a little more to have the bricks delivered on a pallet. In a routine delivery, they will be dumped in a pile, an action than can result in considerable breakage.

Make sure your supplier has a stock of bricks in the type you're using in the event you need an additional supply. (You might decide to extend your wall farther than the initial plan, for example.) You may find that bricks can vary as much as half an inch in the standard dimensions from one area to another. Get all your bricks from the same supplier and preferably from the same batch.

EXPOSED AGGREGATE concrete pool deck is separated from planting area by low wall built with used brick.

— STURDY BRICK retainer supports terrace of bricks set in mortar; terrace is at floor level of house.

BRICK WALLS 65

STAGGERED BRICKS create baffled screen to give privacy to a window close to the entry walk.

SWIMMING POOL COPING tops low brick wall to form comfortable, waterproof seat; wall retains slope.

CURVING SEAT WALL is made of double thickness of bricks; it's topped with strong, interlocking bond.

How to Build a Brick Wall

Laying up a brick wall is exacting work. Because of this, plan carefully and don't rush. A flaw in workmanship will not only be visible but, more importantly, will weaken the wall. Choose your pattern, lay your foundation, and work from the ground up and from the corners (or ends) inward.

Your choice of patterns

Patterns, or bonds, are based on how the bricks are used in the wall. Four commonly used patterns are illustrated here. Bricks are referred to structurally, according to their positioning in the wall: a brick used lengthwise is a stretcher; a brick used across the width of the wall is a header. The pattern you choose will depend largely on the required thickness of the wall. This, in turn, is based partially on the desired height.

```
Running bond        Flemish bond

English bond        Stack bond
```

CHOOSE from variety of wall patterns based on design and required strength for masonry block walls.

Thickness. You can build a wall 4 inches thick (the width of one brick), provided it is not higher than 2 feet. A running bond of stretchers is best suited for this type of wall. An exception to this height limit is the curving serpentine wall that forms its own support. (To lay out a serpentine wall, set out a stake, attach a string, and use it like a compass to mark the arc of the serpentine.)

Higher walls or retaining walls are better if they are made 8 or more inches thick with built-in steel reinforcements. Flemish bond and English bond are strong bonds because both have header bricks that reach across the double thickness of the wall. A variation of the Flemish bond is known as rolok (rowlock) bond—header and stretcher bricks set on their sides. This provides strength and saves bricks.

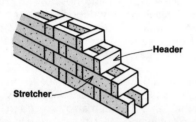

ROLOK BOND constructed of header and stretcher bricks set on sides gives strength and saves bricks.

Various bonds can be combined for necessary strength. A stack bond, not a particularly strong bond in itself, may be backed with a thickness of running bond to add needed reinforcement and stability.

Reinforcing brick masonry

A variety of methods may be employed to reinforce walls over 2 feet high or retaining walls. A common method is to lay the wall up in a double thickness of interlocking bonds, such as described above, or to cap the top with a zone of header. You can also reinforce with steel or additional masonry structures.

Steel reinforcing. A good way to reinforce your wall with steel is to lay the wall up in two separate tiers. Then pour grout (thin soupy mortar) into the hollow interior of the wall. Allow the grout to stiffen slightly, then insert steel reinforcing rods into the grout down to the foundation. You can also set the steel rods into the concrete foundation before you lay up the wall.

Another method is to insert steel beams in the concrete foundation so that they will project upward at the back of the wall. Then run bolts through holes in the beam into the mortar joints of the wall.

Masonry reinforcing. If the wall is going to be fairly long, it should be reinforced every 12 feet with a pilaster or brick pier. For extra strength, and especially if the wall will be 6 feet or more, combine the pilaster reinforcement with the steel rods as described above.

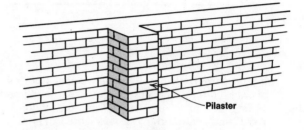

PILASTER locked into brick wall by overlapping bond provides needed reinforcement for long wall.

Building the foundation

For stability, a masonry wall should be laid on a 12-inch poured concrete foundation. But for very low walls, up to a foot in height, the foundation may be bricks laid side by side in mortar on a bed of sand. The foundation should be twice the width of the wall. If you plan to use double bonds, or pilasters, the foundations will have to be wide enough to accommodate them.

To pour a concrete foundation, stake out the length and width of the foundation and dig a trench somewhat wider than the foundation. The surface of the foundation should be low enough so that the first course of bricks is laid below the surface of the ground. (Follow the procedures outlined in the chapter on concrete paving on page 29.)When the foundation is in place and has been allowed to set for at least 24 hours, you can begin laying the bricks.

Laying bricks

Bricks should be damp but not wet when they are laid. If they are too wet, they will weaken the

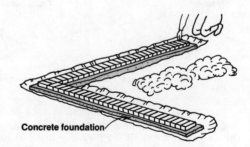

LAY TEST COURSE of bricks out on foundation to check fit before mortaring bricks in place.

mortar, causing it to run, and the bricks will slip in the mortar bed. Wet the bricks down with a fine spray about four hours before you plan to use them.

Test your course. Before you begin mortaring the bricks in place, lay an unmortared course on the foundation to test for fit, leaving a ½-inch opening between bricks for the mortar joint. Try to

---

MIXING MORTAR

If your garden plans call for permanent flooring or wall building, chances are good that you'll be adding mortar-mixing to your mason's talents. You'll find that mortar is a very gifted material; it does much more than bind or hold together materials — bricks, stones, or blocks — to form a masonry structure.

For one thing, mortar cushions and takes up irregularities in the surfaces of the building materials. This helps to equalize weight-exerted pressure throughout a wall and other structures. Then, too, when properly mixed, mortar aids in waterproofing the job. And finished mortar joints between the materials can often become a decorative element when the job is finished.

Don't be surprised to find there are almost as many mortar formulas as there are materials to build with. Building with stone, for example, generally calls for a richer mortar than is required for building with brick. For most jobs, though, a good all-around mix consists of 3 parts portland cement, 1 part fireclay or hydrated lime, and 9 parts sand. The cement binds together the sand with the building units into a continuous structure; lime or fireclay is added to help bind and

add plasticity (workability) to the mixture; sand helps minimize shrinkage and cracking that would occur if cementitious materials alone were used as mortar.

The sand you use should be clean and sharp, consisting of angular particles. When it is wet and you squeeze it, it should not bind together, nor should it leave a slimy deposit in your hand. Water, too, should be clean — and as free as possible from alkalies, salts, acids, and organic substances that may affect the time it takes the mortar to set.

You can use almost any flat surface for mixing mortar: a small platform made of 1 x 4s laid close together, an old wheelbarrow, wooden box, or square of plywood. Mix the mortar in small batches so that it won't dry out. Use up the batch within an hour. If it starts to dry out while you're working, freshen it with a small amount of water.

Remember that the mixing proportions will vary depending upon the building material you're using. To keep these proportions of ingredients accurate, take a few minutes to construct a 1 cu. ft. box. This small chore will save you time and expense by taking the guesswork out of your measuring.

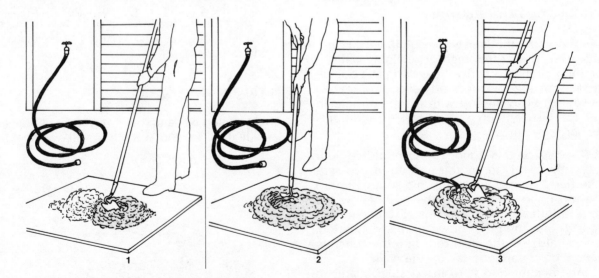

MIX MORTAR on wooden platform, in wheelbarrow, in wooden box, or on square of plywood; 1) place ingredients in a dry state in mixing container or on wooden surface; 2) mix ingredients in dry state and scoop out hollow; 3) add water slowly and mix thoroughly with the ingredients until mortar binds them together in workable mix.

space them out to an even number of bricks. If you're satisfied with the fit, you can begin mortaring them in place.

Building corners. Build up the corners first, working inward from there. To start with, mix a small batch of mortar—enough for about 50 bricks (see box on mixing mortar, opposite page); 1 shovel of cement; 4½ shovelfuls of sand; and ½ shovelful of fireclay or lime.

Slice a trowel full of mortar from the board and spread it ½ inch thick on the foundation where you plan to set the first corner brick. Roughen the mortar bed with the tip of the trowel and press the first brick in place, making sure it is level across the length and the width. Trim off the excess mortar and butter it onto the end of the next brick to be placed on the end that will face the previous brick. When you get the hang of this operation, you begin putting down enough mortar for four to five bricks at a time.

Build up the corners (or ends) in a step-like fashion with the bricks dovetailed so that each overlaps the joint of the previous course. Check continuously with a spirit level and a carpenter's square to make sure the bricks are level and wall plumb. Adjust any brick not in line by tapping it with the handle of the trowel before the mortar sets. If the wall is built with two tiers, build them up at the same time.

Extending the wall. When the corners are built up, run a guide line between the two corners. Attach the line with a nail pushed into the corner mortar joints. The line will act as a guide for each fill-in course and should be moved up with each course. Fill in the holes left by the nails.

Put down a mortar bed ½ inch thick and long enough for about five bricks and furrow it with the end of the trowel. Butter the end of a brick that will face the previous brick and set it in place; tap it until the top is level with the guide line. The last "closure brick" laid to complete the course should be buttered on both sides, along with the two bricks flanking it. Lower the

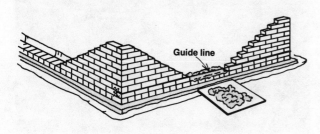

EXTEND WALL from ends or corners inward, one course at a time. Set bricks flush with guide line.

brick in place and tap it level. Follow these same building principles regardless of the bond you use for your wall.

If a brick should slip out of place before the mortar has set, scrape out the mortar bed, replace with fresh mortar, then reseat the brick.

Finishing mortar joints

Mortar joints should be dressed and finished while the mortar is still wet. Trim off any loose or extruding bits and smooth off the joints—the vertical one first, then the horizontal ones. Use any of the four basic joints shown below. Smoothing or "striking" the joints creates a finished appearance and helps bond the bricks together, sealing the wall against moisture.

You can purchase special jointing tools to achieve these finishes, but other instruments may be substituted. The common concave joint can be formed with a wooden dowel. Form the v-shaped joint with the tip of a trowel or the corner of a piece of wood. The other joints may be formed by using the edge of the trowel tipped up or down.

Of these joints, the struck joint is least recommended because it does not shed water readily. The weathered joint provides the best water shedding capability.

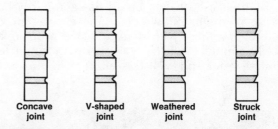

| Concave joint | V-shaped joint | Weathered joint | Struck joint |

FOUR POPULAR ways to strike mortar joints. Concave and weathered joints are most functional in walls.

Cleaning up

Plan to keep the wall moist for a few days. After two or three weeks, you can clean off any mortar stains or traces of white efflorescence (white stains produced by excess water rising to the surface of the brick) that often appears on fresh brick work. Use a 9 parts water and 1 part muriatic acid solution and apply with an old rag. Be sure to wear rubber gloves and safety goggles for protection against the acid. Scrub the surface with a stiff brush if necessary. Wash the wall thoroughly with a hose. If the solution comes in contact with your skin, rinse thoroughly and wash with solution of bicarbonate of soda.

CONCRETE BLOCK WALLS

For strength, economy, and speed in wall construction, concrete blocks are hard to beat. Though concrete blocks are sometimes considered dull and grey, they actually afford more possibilities for creativity in wall construction than perhaps any other building material available. When you visit any building supply yard, you soon realize that concrete blocks head the fashion list in the construction industry. The basic blocks themselves can be utilized in many ways, but in addition you can choose many different kinds of decorative blocks. In fact, shopping for them can present all the problems of deciding on that new pair of shoes.

Building with concrete blocks: pro and con

Concrete blocks provide sturdy wall construction, partly because of their broad size and partly because of their hollow cores. These cores make reinforcing them with steel rods and grout relatively simple. And the wall seems to come together faster and with less effort than one built of clay bricks because of the large size of the standard block (8 by 8 by 16 inches).

In the hands of an imaginative designer, conventional blocks can be arranged to make an unusual and attractive wall. Try combining standard concrete blocks with the smaller sizes available to give the effect of fitted stonework. A concrete block surface lends itself to a variety of surface finishes, such as stuccoing or coloring with masonry paint.

Concrete blocks add a strong structural element to the garden, standing out boldly among the plantings and adding a decorative texture to the landscaping.

Concrete block construction does have some drawbacks. The blocks are not visually well-suited for low walls because of their size. A 2-foot wall of concrete blocks can look oddly out of scale. Handling concrete blocks is a more difficult chore than building with bricks—a standard block weighs about 45 or 50 pounds. Because of their weight, concrete blocks must always be laid on a concrete foundation. They should not be laid directly on soil as you lay a low brick wall.

CONCRETE block retaining wall makes possible this outdoor living area of decks and garden pools.

TALL CONCRETE block wall has one course of blocks set on end; wall is reinforced with steel rods.

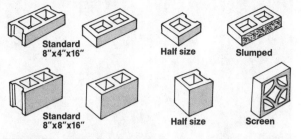

COMMON STYLES and sizes of concrete blocks.

Kinds of concrete blocks

Basically, two types of concrete blocks are used for building walls: heavyweight and lightweight. The former, made of the same materials as ordinary concrete, is called "standard aggregate block" or simply "concrete block." An 8 by 8 by 16-inch standard aggregate block offers greater resistance to moisture and is usually the most economical. Lightweight blocks, made from such aggregates as volcanic scoria, expanded clay or shale, cinders, or occasionally pumice, weighs only about 25 to 30 pounds, has a more porous texture, and offers better insulation.

Though a large variety in styles and sizes is available, you can get by with only one or two sizes: for example, the 8 by 8 by 16-inch blocks and the 4 by 8 by 16-inch blocks. These blocks are made with either two or three cell holes. The three-cell variety is more common in the East, the two-cell type in the West.

Decorative block. Like a fashion guide that shows clothes for all occasions, manufacturers' catalogs demonstrate the versatility of concrete blocks.

Textured blocks are available that look like rough-cut stones or flagstone slabs. You can also build with a variety called slump block or slump stone. Made of high grade concrete, slump block is released from the mold before setting completely and allowed to sag or slump, acquiring individual contours.

A wall made of slump block looks as though it were constructed of weathered stones. Another variety, called shadow block, has subtle ridges on its face that cause interesting and varying shadow patterns to play across during various times of the day.

The most striking of the decorative blocks is screen block. In this, each unit is a frame surrounding a screen or pleasingly contoured hollow. Countless styles are available, and new ones continually are being designed. You can combine various styles of screen block into patterns of your own creation. But guard against making them too busy.

As the name suggests, screen block is designed for building a screen rather than a solid wall. With it, you can't achieve the privacy or protection from the elements that solid blocks afford.

Buying blocks

Plan your wall and estimate the number of blocks you'll need. For the standard block, your dealer will have plenty in stock. If you're using decorative blocks, make certain he has enough in the style you have chosen to complete the wall.

GRILLE OF CONCRETE blocks is built by alternating 2-core blocks laid on sides and set at right angles.

ALTERNATING COURSES of standard and half size concrete blocks add variety and design to long block wall.

SCREEN BLOCK grille is attractive background for desert plants; choose from many decorative blocks.

DIAMOND-SHAPED decorative block mortared on end form this tall screen which provides privacy.

IRREGULAR CONTOURS of slump block give pattern and texture to massive, solid block wall.

SLUMP BLOCK, mortared together vertically, forms graceful-curving wall along driveway.

How to Build a Block Wall

The size and weight of concrete blocks allow you to build most garden walls using only one thickness of the block. But because they are quite heavy, you will have to lay a sturdy concrete foundation twice the width of the wall, using the same procedure as that outlined for concrete paving on page 29.

Patterns to use

A number of the patterns employed in brick construction (see page 66) can be used equally effectively in building with concrete blocks. Since concrete block walls frequently don't require double tiers, you'll have to use a combination of half blocks and full blocks to duplicate some of the favorite brick patterns.

The most commonly used bond for concrete blocks is the running bond. The advantage of this is that the cells of the blocks line up vertically, readily accommodating steel reinforcing rods and grout.

The stack bond is another pattern well-suited for concrete block construction, although it does not have the same lateral strength as the running

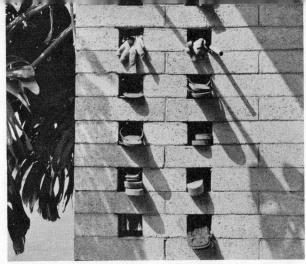

HOLES IN CONCRETE BLOCK wall, created by skipping stretcher blocks, make excellent storage spaces.

SCREEN of decorative concrete blocks offers much privacy; effectively blocks view from outside.

VIEW FROM INSIDE same screen as shown left shows good view outward; screen does not block out light.

bond and should be limited to walls no higher than 4 feet.

Laying concrete blocks

To get a more accurate idea of how to lay the foundation, you'll find it helpful to lay out an unmortared course of the blocks even before you dig. Whenever possible, plan your wall so as to avoid having to cut the blocks. The larger blocks are much more difficult to cut to size than clay bricks, and cut blocks can detract from the appearance of the wall.

Once the foundation is down, you have your choice of laying the wall on a dry foundation or a wet foundation.

Dry foundation. For the beginning builder, it is easier to lay the wall up on a dry, hardened foundation. Building the wall and the foundation are treated as separate steps; the footing is poured in sections or all at one time and then allowed to set before the blocks are laid on top of it.

To check on the fit, lay a course of unmortared blocks out on the foundation, allowing for ⅜-inch mortar joints. Mark the positions of the blocks after making any necessary adjustments and set the blocks aside.

Mix a small batch of mortar to start with (for the procedure see page 68). Because the concrete blocks exert more pressure on the mortar bed, keep the mortar on the stiff side. A soupy mortar would be squeezed out of the joints.

Clean off the concrete foundation and wet it down. Trowel a 2-inch deep bed of mortar the full size of the block directly onto the foundation. Set the two corner end blocks first, making certain they are level along the length as well as across the width. Press the block down to an accurate ⅜-inch joint between the block and the foundation. Since you'll level the rest of the course against the corner block, make sure it is properly set and leveled.

Unless the wall actually has corners, you will be using alternate full and half size blocks to lay up the ends evenly. After placing the corner blocks at both ends, run out the rest of the course.

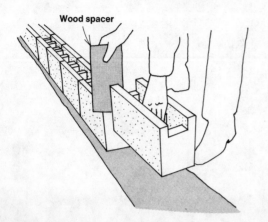

RUN TEST COURSE of concrete blocks on foundation. Use wood spacer to maintain even joints.

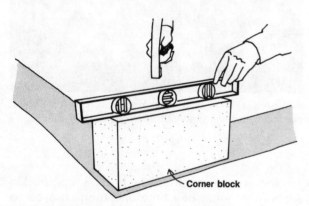

SET CORNER BLOCKS first; use spirit level; tap block with hammer to set level on mortar bed.

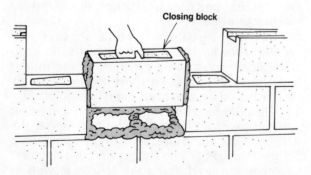

CLOSE COURSE with closure block buttered with mortar on all edges for proper fit and strong joints.

Blocks placed on top of other blocks should be mortared on the outside edges only. To place the final closing block, butter all the edges that will make contact and lower the block into place.

Wet foundation. Laying blocks on a foundation that is still plastic will produce a sturdier wall because the first course is embedded in the concrete. This method is more difficult than laying blocks on a dry foundation.

Because the entire foundation must be cast at one time, order transit mix that can be poured in one operation.

To check the positioning of the blocks, run out the bond on the floor of the foundation trench. If you are satisfied with the fit, pour the foundation and strike the surface smooth and level. When the concrete has set up to about the consistency of mortar, seat the blocks into it about 2 inches deep.

To make the vertical joints, butter the inside end of each block from the second one on, using mortar as described earlier.

For succeeding courses, the process is the same as that for the wall set on the dry foundation. Using a mason's line as a guide, start at the corners with squared corner blocks and lay up the rest of the course.

Sealing the wall

The top course of a concrete block wall must be sealed to prevent water penetration. One of the simplest methods is to fill the hollow cores with grout. To avoid filling the cells all the way to the bottom, lay a strip of builder's paper along the top of the next to last course. Lay the top course and then pour the grout mix into the cells.

Another method used to seal the top of the wall is to cap the wall with a solid block. (Special concrete cap blocks are made specifically for this purpose.) Coping may be of wood, bricks, or concrete block veneer. For masonry coping, trowel a layer of pure sand and cement grout along the top of the wall and place the coping. If you plan to cap the wall with wood—to make a seat wall, for example—place the builder's paper along the top of the next to last block and lay the top block as before. Fill the cells with mortar and put threaded bolts centered in the cells containing the mortar. The wood coping can then be drilled and bolted down. Bolts should be recessed below the surface of the coping.

Reinforcing block walls

Vertical reinforcing rods are required if the wall is over 3 feet in height or must withstand strong wind or soil pressures. Set the rods firmly in the concrete foundation, spaced to match the holes in the blocks and to conform to local building codes.

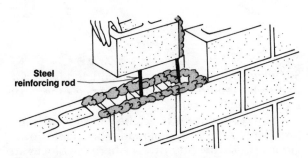

PLACE ONE CELL of concrete block over vertical reinforcing rod. Butter all four edges around rod.

When laying the wall, the cells through which the rods pass should have all four sides buttered. When the wall has been laid, fill the cells containing the rods with a mixture of grout and gravel. In most communities, grout can be poured only to a height of four feet in a single hand pour. If the wall is higher, a single pour must be done with a high-lift grout machine. If hand-poured, a lateral reinforcement must be set (and inspected) at four feet and a higher pour made later.

For horizontal reinforcement, steel rods can be placed in a course of specially grooved bond blocks as illustrated below.

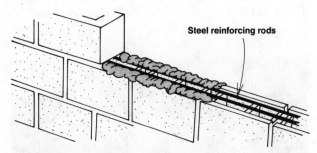

STEEL RODS between courses provide horizontal reinforcement for long concrete block walls.

As with brick walls, long concrete walls should be built with reinforcing pilasters every 15 to 20 feet. The pilaster may be "locked" into the wall bond with an alternating header and stretcher pattern.

Laying concrete block grilles

Ordinary concrete blocks laid on their sides have been employed very successfully for grilles. The standard concrete block grilles may be laid up the same way as the closed wall, in either the running or stacked bond. Combinations of patterns are endless.

Laying screen blocks

A wall of decorative screen blocks will not be as strong as a wall of standard blocks. For the most pleasant results, the screen blocks are best laid up in a stacked or a running bond pattern. Too many combinations of patterns or variations within the bonds could make the wall appear too busy.

Many professional builders prefer to lay up a wall of screen blocks with an epoxy mortar. This mortar sets up incredibly hard, often making the joints between the blocks stronger than the blocks themselves. The drawbacks of this material are that it is expensive (up to 40% more costly) and much more difficult to work with than ordinary mortar. The mortar sets up fast and must be used quickly.

Finishing mortar joints

Mortar joints between concrete blocks are finished in much the same way as brick joints (see page 69), except that the jointing tool should be considerably longer (about 20 to 22 inches) to reach across the irregularities in the horizontal joints of the blocks. The ends should be bent to prevent them from stabbing into the mortar.

The most common joints used for concrete block walls are the concave joint and the V-shaped joint.

Coloring block walls

Portland cement paint provides the most lasting and waterproof surface coloring. It should not be applied sooner than 24 hours after the wall is finished and not at all if the temperature is below 40° or if there is frost on the masonry. If at all possible, the paint should be applied in the shade. The wall should be given two coats.

Clean the wall surface of oil, grease, dirt, and dust. Remove efflorescence with a muriatic acid solution and fill in the small cracks with the thick paste of the cement paint. Mix the paint according to the manufacturer's instructions.

Dampen the wall with the garden hose and wait until the surface water is no longer visible. Then brush on the paint with a coarse-bristled brush, painting the mortar joints first. Keep the surface wet for 24 hours. Wait 24 hours before applying the next coat and keep the finish coat damp for 48 hours.

ADOBE WALLS

You can bring a bit of the old Southwest into the garden when you build your garden walls of adobe. Today's adobe blocks, fortified with asphalt stabilizer, offer strong, long-lasting construction. They are much improved over their ancient predecessors which, over the period of years, crumbled and decomposed because of the seasonal wetting and drying processes. The rugged, earthen-like blocks can bring welcome relief to a garden caught up in this age of plastic and concrete.

Building with adobe: pro and con

Adobe blocks add a friendly charm to the landscape when they are used to enclose a free and open garden plan with large trees and open vistas. The generous size of the blocks seems in scale with the trees and the spreading space; the texture of the wall can be admired even from a distance. Their natural color harmonizes with the warm tones of redwood or cedar house trim, or the wall may be painted to match more formal exteriors. Adobe is an efficient insulator and helps keep the patio cool during warm summer months.

Stabilized adobe is waterproof and, left in its natural state, requires little maintenance other than an occasional dusting.

Still, your choice of adobe can present some problems. Adobe is not readily available in regions outside the Southwest where the blocks are manufactured. Adobe is only economical when the manufacturing plant is nearby. Added shipping costs for long hauls can make the cost of the material prohibitive.

Adobe blocks are considerably more cumbersome to handle than clay bricks: the blocks can weigh up to 45 pounds, and laying them is quite a chore, especially once the wall reaches chest height. Try to avoid using adobe in a confined space where the large blocks will look overpowering and out of scale.

Kinds of blocks

Adobe blocks used for wall work are available in a variety of sizes. The common size is 4 inches deep by 16 inches long, with width varying in size from 3½ to 12 inches. A 4 by 7½ by 16-inch block is the equivalent of 4 or 5 clay bricks.

ADOBE BLOCKS combined with iron grillwork to form secure, yet open feeling to front yard wall.

TWO-FOOT WIDE adobe block flower bed encloses all-adobe patio; contains colorful marigolds, zinnias.

Laying adobe blocks

The method employed to lay up an adobe block wall is similar to that used in laying clay bricks or concrete blocks as described in the previous chapters.

Because of their weight, adobe blocks require a sturdy foundation (for the building technique see the chapter on concrete paving, page 22). Since heavy blocks are not as shock resistant as clay bricks, their foundation must be stable enough not to yield to soil movement. Tensile strength is added to the wall by lateral steel reinforcing rods (two ¼-inch rods every third course works well). Steel reinforcements are also essential at corners to anchor the two stretches of wall to each other.

If you need to cut the blocks for fit, you can use the same procedure outlined for cutting bricks on page 15.

Mortar. Adobe construction requires a leaner mortar mixture than is required for clay bricks. Use 1 part cement, 2 parts soil (the same as in the bricks), 3 parts sand, and 1¼ gallons of stabilizer per sack of cement (or about 1 pt. per shovelful).

How to order

Inquire about purchasing adobe blocks at a local building materials yard. If they don't carry a big supply, they can order additional blocks directly from the manufacturer. It is possible to order directly from the manufacturing site, but if you require a long haul, the manufacturer may send the bricks at regular freight truck rates or will ask you to make arrangements with a hauling contractor.

How much to buy. Using the 4 by 7½ by 16-inch block, *for 100 square feet of wall, with ¾-inch mortar joints, you will need the following materials:*

Adobe blocks	200
Sand (cubic feet)	9
Portland cement (sacks)	3½
Stabilizer (gallons)	5½

The amount of reinforcing steel will depend upon the height and length of the wall. On an average, 100 square feet of 6-foot wall requires ten ¼-inch pencil rod pieces 16.7 feet long (28 pounds).

Finishing adobe

Adobe block walls can be painted to achieve a variety of effects. The new cold water latex or vinyl cold water paints are proving very satisfactory on good quality adobe. If you use ordinary oil masonry paints, the asphalt stabilizer might bleed through.

If you're building or veneering with good-quality adobe block, it's a good idea not to paint over the natural earth color until you see how it looks with the landscaping. Once you paint over the bricks, their original rustic appearance is gone — the texture will never look the same if you later remove the paint.

Stucco colors (oxides) may be added to the cement mortar to obtain any color desired.

PAINTED ADOBE WALL encloses pleasant patio entrance to adobe house; seems integral part of structure.

BURNT ADOBE BLOCK wall extends from house to enclose pool and patio; native desert is right outside.

STONE WALLS

Stone masonry is an ancient craft. The pueblos built by prehistoric civilizations in the Southwest were elaborate apartment complexes of carefully fitted stone cut and shaped from surrounding rock formations. Large sections of these structures still stand today, evidence of the durability of stone masonry.

Building with stone: pro and con

Natural stone masonry will never look out of place in the garden. Stones may be set dry or in mortar. By carefully fitting and stacking them, you can build a low garden wall, retaining wall, or raised bed; the irregular shapes of the stones hold them securely in place. A nice feature about stone walls laid up without mortar is that you can plant in the joints, adding to the natural look.

Stone does have some weaknesses as a building material. Compared to brick work, stonecraft is difficult. Stones are heavier to handle, and uneven surfaces make plumb lines difficult to achieve. Handling the larger stones is especially hard work; setting them can be awkward, for stones can squash the mortar bed or jam in wrong positions.

NATURAL ROUNDED STONES were laid dry to form double retaining wall. Plants in cracks help bind wall.

Another factor to consider is cost. If you have a ready supply of stones in your yard, you invest only your time and your good strong back. If, however, you have to order the stones from a supplier, a small garden project can become a substantial financial investment.

Kinds of stone

If you're buying the stone, you'll find that local stone dealers offer a considerable variety from which to choose. Stones that are gathered or quarried locally will probably look more at home in your garden than stones shipped in from other areas.

Almost any kind of stone that is available in quantity may be used, though you'll find the stratified variety, such as limestone, shale, and sandstone, the easiest stones to trim and face. Formed of layers of solidified soil deposits, stratified stones are easy to split and chip, but this same quality limits their value in severely cold climates: they absorb moisture that can split the stone when it freezes. Solid mortaring, followed by an application of a good masonry sealer, will usually prevent this by keeping the moisture out.

The most durable stones are the granites and basalts. Their very toughness, though, makes them an obstinate material to handle — hard to break and chip. Because they absorb very little water from the mortar, their drying time is lengthened, making them slow to lay up.

Types of stone work

Stonework is divided into two broad classes: rubble and ashlar. Rubble masonry is composed of uncut stones, fitted into a structure in their natural state. Ashlar masonry is built with cut stones laid up in fairly regular courses much like brick, concrete block, or slumpstone (see page 80).

For rubble masonry, unstratified stones are more commonly used because they are difficult and expensive to cut. Ashlar walls are laid up primarily with stratified stones because they are easy to cut and dress.

Although ashlar walls are easier to construct,

(Continued on page 80)

CUT AND DRESSED stones can be laid up in mortar in much the same way as bricks. Sandstone laid up in ashlar pattern; care was taken to prevent staining stones with mortar. Needs water seal.

CAREFULLY FITTED sandstone in ashlar bond wall gains strength with placement of long stretcher stones.

CREEPING ROSEMARY cascades down wall of stones set in mortar. Planting enhances stone walls.

GRANITE BLOCKS set in mortar in a formal diagonal pattern. At the top is a clipped pyracantha hedge.

cut and dressed stones are frequently more expensive.

How to buy stones

Estimates for the amount of stone required for a wall are generally figured in tons. Required tonnage differs from dressed to undressed stone. One ton of quarry, field, and other types of undressed stone will cover 25 to 40 square feet of wall with an average thickness of 1 foot of stone. One ton of dressed stone will cover about 55 square feet of wall surface with an average thickness of 6 inches of stone.

If you provide your dealer with the cubic wall area of your wall, he will compute the quantity

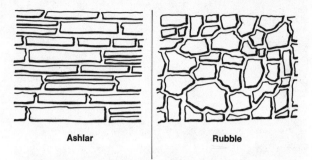

Ashlar   *Rubble*

ASHLAR MASONRY made with cut and shaped stones. RUBBLE MASONRY is composed of uncut stones.

needed for the job. Some dealers sell stone by the cubic yard.

It is advisable to inspect the stone before you decide to buy it. Stones should match in

MASSIVE ROUNDED stones laid up in mortar to form retaining wall and raised bed along sparkling pool.

RUBBLE STONE retaining wall and flagstone bench hold slope covered with gray mounding cushion bush.

DRY STONE WALL with crevice planting borders informal cobblestone path; wall holds soil for raised bed.

color and texture and should have a good mix of sizes. You'll need a variety of sizes to give the wall necessary strength. A good rule to remember: the face area of the larger stones laid should not be more than five or six times the face area of the smaller stones.

If you're ordering cut and dressed stones, order specific thicknesses and designate the upper and lower limits of length. Remember, larger stones can be cut if necessary.

Building key: careful fitting

The key to building an attractive stone wall is careful fitting, whether the wall is laid up with or without mortar. From a standpoint of both appearance and structural strength, stones should be fit to create a balance throughout the wall between large and small stones — this will ensure a strong bond. Two precautions: be sure to have enough stones on hand, and provide the wall with a sturdy foundation.

Stonework patterns

You can exercise a good deal of freedom when designing stonework. Properly placed stone should produce harmonious and pleasing patterns in which there is variety within a basically harmonious composition. The finished structure should appear to be a unit rather than a conglomeration of rocks.

CLOSE UP VIEW of dry stone wall reveals plant life in soil-filled cracks and crevices; plants bond stones.

RUBBLE WALL of native rocks lines generous tree well; keeps soil around tree at original level.

LOW WALL of stones set in concrete is capped with wooden planks and edging to form seat wall by pool.

Lay the stones as they would lie naturally on the ground — rarely on end or in unnatural positions. Avoid stacking the courses in continuous wavy joints, for the resulting "lightning bolt" pattern destroys the feeling of strength and solidity inherent in good stonework.

Use most of the larger stones in the lower courses and most of the smaller stones in the upper courses. Avoid setting stones of the same shape and size together. At the same time, guard against setting in a long, narrow stone that is much larger than adjacent rocks.

To strengthen a stone wall, bond it by overlapping two small stones with one large one.

The foundation

For strength, the foundation of any masonry structure should be built of concrete or stone set below the ground (below frost level). Generally 10 to 12 inches for a 3-foot wall is sufficient depth for the trench in which the foundation is laid. The bottom of the trench should be firm ground; avoid placing the foundation on filled ground. The surface of the foundation should be just below the surface of the ground. Set the stones in mortar directly on the foundation.

Concrete is generally considered the easiest foundation to install. Follow the procedures outlined for concrete paving on page 29.

Setting stones in mortar

Have plenty of stones handy, offering a good selection of shapes and sizes, and place them where you can reach them as you work. Clean the stones thoroughly, removing all dirt and lichen from the surface to be mortared. Avoid brushing them with an iron-bristle brush, for doing this may produce latent rust stains. If you need to use water to clean the stones, let them dry before you mortar them.

Assemble the tools you will need: a heavy hammer, a pointed trowel, a cold chisel or brick set, a spirit level, some string or mason's line (see guide on tools, page 13). You'll also need materials to mix mortar.

Mortar for stonework. A wall of stone requires much more mortar than a brick or concrete block wall. Because of the larger joints and voids in stonework, as much as ⅓ of a fieldstone or riverstone wall may be mortar. To estimate the amount you'll need to mix, lay up a small section of the wall and, based on the mortar used, compute the amount for the rest of the wall. The formula recommended for stonework is richer than that required for brick: 1 part cement; 3 parts sand;

½ part fireclay. Don't substitute hydrated lime for fireclay because it is likely to discolor the stones.

Construction. Choose larger flat stones to mortar in a stable position directly on the foundation. String up a guide line, keeping the face of the structure flush by selecting and setting stones so they don't jut beyond the face of the wall. To keep the pattern interesting and to insure good bonding, try several stones in a section of the wall before mortaring them in place.

Give the wall a proper slope by nailing three boards together in a triangle (see illustration) and holding it against the face of the wall. The outside

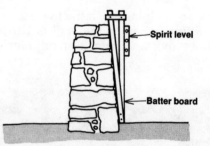

CHECK SLOPE with batter board set against sides of wall; outside edge should be plumb.

edge should be plumb. On the average, a freestanding wall should slope 1 inch for every 2 feet of height; a retaining wall, 1 inch for every foot of height.

Bond the wall by overlapping the vertical joints at every course. To strengthen the wall transversely, put in headers (stones set with their long dimensions at right angles to the face of the wall).

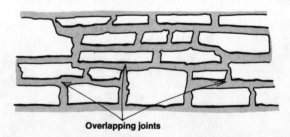

OVERLAP VERTICAL JOINTS at every course for strong bond. Place header stones across width of wall.

Use enough mortar to fill the joints completely. Fill empty spaces in the interior of the wall with small chips of stone and mortar. If the wall is not fitting properly and you have to shift stones already set in mortar, lift the stones out and scrape off all the mortar, replacing it with a fresh layer.

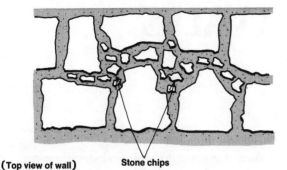

(Top view of wall)   **Stone chips**

FILL INTERIOR SPACES in wall with small stone chips and mortar; chips help bind, and save mortar.

After you have laid a section, rake out the joints on the facing before the mortar sets. Rake them out using a dull pointed stick, ½ to ¾ of an inch; the deeper the rake, the better the shadow effect. Thoroughly brush off all the excess mortar with an old broom before the mortar sets too firmly.

RAKE OUT MORTAR around stone edges with dull, pointed stick. Brush off excess mortar after raking.

Veneering with stone

Another way to lay a stone wall is to set stone veneer against a tier of bricks or concrete blocks. Concrete blocks provide greater strength. Stone veneer need not be set on a slope like a free-standing, all-stone wall.

Generally, cut stone is preferred for veneer work because the squared surfaces are more easily aligned with the flat planes of the bricks or blocks. Uncut stones may be used but can produce an unpleasant knobby texture because they are set against a hard flat surface that forces uneven facing. The easiest type of veneer to apply is dressed sandstone, available from most building supply yards in strips 3 or 4 inches wide, 1½ inches high, and from 6 to 24 inches in length. Colors range from brick red through buff to yellow. Sandstone can be cut and laid up like brick but must be laid dry.

Build the supporting block wall first (for technique see the chapter on concrete block construction, page 70). Mortar the veneer solidly to the block wall. Remove mortar stains from the stone immediately with a damp rag. To remove stains after the mortar is dry, use an alkali soap solution (2 bars to 1 gallon of water). Apply this solution to wet stone, scrub, and rinse thoroughly. Remove soap stains with vinegar. Do not use muriatic acid on limestone or sandstone.

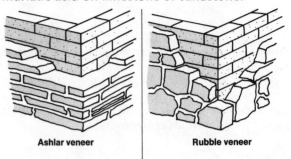

Ashlar veneer   **Rubble veneer**

SET VENEER with ashlar or rubble masonry; mortar stone veneer to masonry block wall.

Building a dry wall

A dry wall, one that is laid up without mortar, is an informal and attractive form of stonework. Stones are carefully fitted together in mosaic fashion so that they hold each other in place by weight and friction.

Free-standing wall. A free-standing wall is built without a deep foundation, even in climates with severe frost. If the wall lists because of winter rains or frost, it is easily repaired in the spring. Although the wall may not have a foundation, it is well to lay up the first course just below the grade of the soil. Hollow out the shallow trench just wide enough to accommodate the first course stones. The snug fit in the trench helps stabilize the base stones and keep them from shifting.

The wall is held together by the bond stones that pass through it. One bond (header) stone should be placed for every 10 square feet of wall surface (about 1 every 3 feet, every third course). Corners and ends should be almost entirely bond construction. Try to get the top course as level as possible; save some of the flat stones for this.

Retaining walls. A solidly constructed dry wall can serve as an effective and distinctive retainer. Place stones so that their back ends tip down into the ground and their front faces slightly upward. This backward slanting provides better resistance against the pressures from the earth behind. The steeper the bank, the greater the slope of the wall.

Use larger stones in the lower parts of the wall. High steep walls require large stones throughout. If you plan to grow plants in the joints of the wall, they should be set in when the wall is built.

CAST CONCRETE WALLS

Whenever strength is essential in construction, man turns to concrete more than to any other material. Cast concrete supports the nation's skyscrapers; concrete dams across rivers provide us with electricity; and concrete is shaped in flowing curves to form freeway overpasses.

Building with concrete: pro and con

Concrete cast in place has some unique advantages over other building materials: strength and workability head the list. Much of the inherent strength of poured concrete results from its single unit construction. Because concrete is poured in a continuous mass, there are no mortar joints to weaken the structure. Internal steel reinforcing is completely surrounded by the concrete and becomes an integral part of the wall.

Concrete may be formed in almost any desired shape to fit the garden plan. Curve it, pour it in a V-shape, or make it square.

With concrete you have your choice of surface finishes. The finish can be smooth, rough textured, or embossed with architectural lines. If you use a coarse grained wood to build the forms, that grain will be stamped onto the surface of the wall.

But the benefits of poured concrete do not come easily. Casting a concrete wall is certainly no Sunday afternoon job for the handyman.

SMOOTH CAST CONCRETE wall contains raised planting bed at one end and encloses firepit on the other.

CURVING CONCRETE WALL contains raised flower bed, lines concrete patio; finish is exposed aggregate.

EXPOSED AGGREGATE makes attractive retaining wall and steps; aggregate exposed before concrete sets.

Even the simplest walls call for carefully fitted and strongly built forms to hold the concrete while it cures. Constructing the forms is the major part of the work and will take up probably ⅔ of the total time required to build the wall. Especially for curved or sharply angled walls, building the forms will be a time-consuming chore.

When the wall you plan to build reaches 6-foot dimensions, you will have to construct platforms, ramps, and staging to accommodate the loaded wheelbarrow that must be wheeled to the top of the forms.

If your wall is going to be on a scale larger than 6 feet or if you need the wall to retain a particularly steep sloping hill, get help from a structural engineer and a reliable contractor. You'll find this assistance well worth the additional investment.

Choosing concrete mix

The type of concrete mix you choose to work with depends on the size of the job. For a simple low wall or raised bed, you can just buy the dry ingredients (cement, sand, and gravel) and mix your own concrete by hand in a wheelbarrow or with a small power mixer (for details see the chapter on concrete paving, page 29). The forms can be easily filled from a wheelbarrow with its wheel resting on the ground or on a plank ramp. Or you can pour concrete directly from the drum of a power mixer if it is positioned nearby.

For more ambitious wall projects, consider the benefits of ready-mix. A ready-mix truck could fill the entire form in one load, provided you have the extra help to assist in placing the plastic concrete. For building sites that are difficult to reach, suppliers of ready-mix have special equipment to pump concrete through a wide hose from the truck over obstacles (such as a fence or part of a house) into the forms. Be sure to check with the concrete plant, for some have requirements for a minimum order.

Some concrete suppliers have special trailers that you can haul behind your car or truck; these carry about 1 cubic yard of ready-mix concrete. The trailers are equipped with either a revolving drum, which mixes the concrete as you drive, or a container carrying plastic concrete already mixed.

RETAINING WALL uses field stones hand set in concrete. Stones set in forms between layers of poured concrete.

WIDE, CAST concrete wall retains soil of raised planting areas; creeper plants cascade over side.

FREE FORM cast concrete wall contains sand box; when kids are grown it can be converted to planting bed.

How to Cast a Concrete Wall

Pouring concrete into the forms is almost anti-climactic after you have spent considerable time and effort building the forms to hold the concrete while it cures. A concrete wall may be poured in sections, but each section must be completed before work stops. A stop-and-go operation could cause air pockets or other faults that would result in weakening of the wall after the concrete has cured.

Building the forms

A wall foundation should be twice the width of the wall, except for a very low wall, and should be laid on firm ground, hard ground, or below the frost level in areas subjected to severe frost. For a low lightweight wall, pour the foundation at the same time as the wall. Taller, heavier walls require pouring a separate foundation keyed on the surface with a slot for stability.

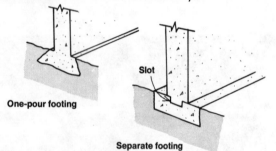

One-pour footing

Slot

Separate footing

POUR wall and foundation in single pour for low walls. Pour separately for higher concrete walls.

Straight walls. Forms should be built with lumber that is free of knotholes, warpage, and other wood flaws that might weaken the form. Plywood gives a smooth finish and also provides good strength. Brace the form with 2 by 4 guide rails

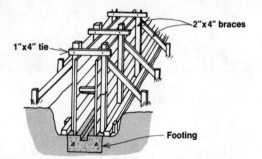

2"x4" braces

1"x4" tie

Footing

CONSTRUCT STRAIGHT FORMS of smooth, knot-free wood; brace form with 2 by 4s set against the ground.

along the bottom and 2 by 4-inch braces, formed in a triangle, against the ground (see illustration).

Curved walls. Use 1 by 4 or 1 by 6-inch lumber, capable of being bent by saw kerfing, to build curved forms. Saw about half way through the broad side of the wood — a power circular saw set at a ½-inch depth is the easiest to use. When the lumber has been cut at regular intervals, taking into account the degree of the curve, soak the wood thoroughly and bend the board towards the cuts. This type of form requires especially secure bracing because of the additional pressures exerted by the bent wood. For low walls, you can use thin plywood that can be bent without saw kerfing.

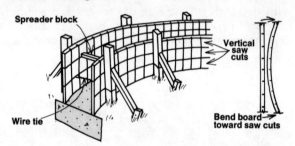

Spreader block

Vertical saw cuts

Wire tie

Bend board toward saw cuts

SAW KERF 1 by 4 or 1 by 6 lumber or bend plywood sheets to form curved concrete wall forms.

Bracing. Opposite posts should be joined with wire ties to help the form resist the pressure of the concrete. The wire is passed from post to post through the joints between the boards of the forms. For additional strength, join the posts over the top of the form with wood ties. Wood ties are especially important when using sheets of plywood for the walls of the forms.

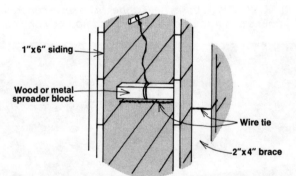

1"x6" siding

Wood or metal spreader block

Wire tie

2"x4" brace

SECURE FORMS: wood ties over top, wire ties through sides. Spreader block keeps sides of form apart.

To keep the forms properly spaced, use 1 by 2-inch spacers cut to wall width and wedged between the walls of the form. Attach a wire to the spacers so you can pull them out just as you place the concrete.

Preparing forms. Keep the inside surfaces of the forms wet for 12 hours before placing the concrete. This causes the wood to swell and seal itself. To make removal of the forms easier, the inside of the walls may be lightly coated with a thin oil or special release agent. If the concrete wall is to be painted later, you will have to thoroughly clean off the coatings from the surface.

Mixing and pouring the concrete

For concrete walls, use a 1 part cement, 2¼ parts sand, and 3 parts gravel, with ⅔ parts water concrete mixture. In regions where winter frosts are a problem, add an air-entraining agent to the concrete mix. This agent causes tiny air bubbles to form in the concrete, allowing it to expand during freezing. (For details on mixing concrete, see the chapter on concrete paving, page 29).

A free-standing wall that runs longer than 35 feet or one that curves or has intersecting corners requires joints that allow for expansion and contraction of the concrete. To create the joints, pour the wall in sections, using a partition as illustrated below. Pour one section, allow it to cure, remove the partition, and pour the next section, and so on.

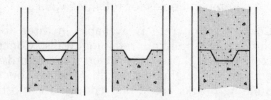

Top view

POUR LONG WALL in sections. Resulting joints let wall expand or contract with changing weather.

Pouring. Pour the concrete in continuous layers 6 to 8 inches deep. Prod and tamp the concrete in place with a 2 by 4 or shovel blade. For a smooth surface, agitate the concrete next to the forms to force the large aggregates away from the surface. A straight-bladed shovel is best suited for placing concrete.

If you have to quit work before a section has been completely poured, cover the top of the concrete with damp sacks and keep them wet until you resume work. Do not discontinue work for more than 30 minutes.

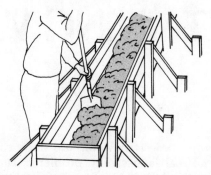

POUR CONCRETE from one end of form in layers of 6 to 8 inches. Tamp concrete in place with shovel.

Capping the wall

A concrete wall may be capped a number of ways to help seal the top surface or to make it more functional.

When the wet concrete reaches the top of the forms, float it with a wooden float and let the water sheen disappear. Then trowel the top with a steel trowel. After the concrete sets, apply a masonry sealer. To make a bench wall with a wood seat, set bolts in the concrete when the mixture is still wet. Then when the concrete has cured, attach the 2 by 6 board to the bolts. Set the bolts in the concrete deep enough so the ends will not project up beyond the surface of the wood. Then nail a 2 by 10-inch board to the 2 by 6 for the final cover.

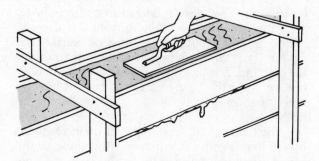

FLOAT CONCRETE at top of form with wooden float. Use steel trowel for final finish.

Curing

Wall forms should not be removed until the concrete has thoroughly hardened. Leave them in place at least 4 days, longer for heavier walls. Keep the top surface of the wall moist while it cures. If it is necessary to remove the forms before the concrete has cured to apply a special finish to the sides, cover the entire wall with sacking or tarpaulins. Keep the covering wet for a week.

RETAINING WALLS

When your home is on a hill, the slope can present some real landscaping problems. How can you keep the hill in place and make the slope useful for landscaping purposes?

It might help to recall one of the ancient wonders of the world. The hanging gardens of Babylon were actually a series of retaining walls and raised beds artfully landscaped with exotic shrubs, vines, and trees. And this principle of the retaining wall has been put to good use throughout the world. You'll see it in the terraced fields along the Rhine and in the rice paddies on Oriental hillsides.

For your own home, plans for a retaining wall will probably not be so ambitious, but you can put the ideas to good use to add three dimensional contours to your hillside garden, at the same time keeping the hill from sliding away.

Types of retaining walls

A retaining wall does what its name implies — it retains, usually a layer of soil ranging from a few feet of raised flower bed to an entire hillside.

Most commonly, retaining walls are used to define a patio, support a terraced surface of a patio, or retain an elevated mound of soil used for planting. A retaining wall may be strictly decorative, as in the case of a raised bed; it may be functional, merely holding a hill; or it may be both.

The type of wall used depends on the amount of soil that is to be retained and the slope of the hill. A gentle slope may be tamed with a single low wall or a series of garden steps that hold the surface soil in place. Low walls or walls of raised beds that do not have to withstand severe soil pressures may be built of loose stones or blocks set on the surface of the soil without being mortared or fastened together.

Steep slopes, on the other hand, require solid walls with sturdy foundations embedded below the surface of the soil. The wall may be one massive concrete wall cast in place or a series of masonry walls dividing up the load-bearing burden. Retaining walls under much pressure are best constructed on specifications from a qualified engineer.

A long, steep slope held by two or three substantial walls will provide a wide range of landscaping possibilities. You can build terraced patios or display a mass of plants, and the plants in the rear will not be hidden by those in front.

Sometimes in the grading of a new homesite, the lay of the land around a tree is disturbed. A tree is difficult to move and may resent modification of its environment, especially a change of soil level or a marked increase in the amount of water it receives. As a result, a tree might be left in a hole or on a raised island of dirt. Both these irregularities may be overcome with retaining walls forming either tree walls or tree wells.

Choice of materials

When choosing a material to build with, consider required strength, secondary use, and appearance. Keep in mind that, regardless of the material used, the wall may be dressed afterwards with a variety of facings — plants that spill over the wall, for example. The following materials are the common choices for retaining walls.

Concrete. Concrete is available in three forms suitable for building retaining walls. Concrete cast in place in pre-constructed forms affords the greatest soil retaining strength. Modular concrete building blocks are strong enough for most retaining situations but require mortaring and steel reinforcing anchored in a solid cast concrete foundation. Slump block is another concrete product quite suitable for retaining walls. For walls 3 feet and higher, lay them up in double tiers with reinforcing rods in the middle embedded in a concrete foundation.

Clay bricks. The use of clay bricks should be limited to low walls and raised beds. Bricks are not as well suited for higher walls because of the small size of the unit and the many mortar joints that would be required. Two-tiered brick walls are more satisfactory.

Natural stone. For versatility, natural stone is hard to beat. For very low walls and raised beds, stones can be laid dry — their uneven surfaces will hold them in place. For higher walls, the stones can be set in mortar. Use uncut stones if you plan to lay up the wall without mortar. Their irregular sizes and shapes afford more stability than the more evenly shaped cut stones. Give stones set in mortar a solid foundation of poured concrete. Stones laid without mortar have the added advantage of allowing plant growth between the joints; this increases soil retention.

Wood. You can choose from a variety of wood products to build your retaining wall. Wood that has been properly treated with fungicidal and insect-repellent preservatives and is securely braced by supports anchored in the soil provides a strong and long-lasting wall. Wood retaining walls built with 2 by 6 or 10-inch boards should be limited to 2 or 3 feet in height and require 4 by 4-inch supporting posts set well into the ground or set below grade in concrete.

Increasingly popular in retaining walls are railroad ties that have been pretreated with preservatives. These can be stacked or set upright in the ground. Their size and bulk make them an effective soil retainer. Not easy to get hold of, utility or telephone poles can also make attractive low walls.

BUILDING WITH BROKEN CONCRETE

BROKEN CONCRETE pieces are ideal for retaining walls. Lay up wall with mortar or stack concrete pieces dry. You can grow plants in the cracks.

If that old, cracked concrete patio or walk must go, why not strike the sledge hammer and a note for ecology at the same time and recycle it? Broken concrete makes a very attractive retaining wall, having the same informal charm as a wall of stone.

Handle broken concrete in the same way as natural stone: lay the wall up dry or set the broken pieces of concrete in mortar.

The work involved in preparing the concrete pieces for wall construction is really no more than is required in removing the old patio or terrace to begin with; the concrete must be broken to facilitate removal. Like stones, the concrete pieces may have to be shaped. This is best done with a 3-pound sledge hammer and a brick set or cold chisel. To break up the patio or walk, it's best to use a heavy, long-handled sledge hammer.

Caution: whenever you're breaking concrete or stone, and especially when using a heavy hammer, wear long sleeves and pants and protect your eyes with plastic goggles. The power and velocity with which a heavy hammer strikes the concrete can send flying small sharp pieces of concrete capable of cutting the skin or piercing an eye. Wear gloves when you handle the jagged chunks.

SERIES OF TERRACED RETAINERS make this garden setting possible. Lower walls are rubble stonework set in mortar; single row of stone contains terraced lawn and raised flower beds farther up the slope.

RUSTIC RAILROAD ties are dove-tailed at joining corner of raised planting bed; forms sturdy retainer.

MASSIVE RETAINER holds small lawn to its steep hillside site; provides level outdoor living area.

NATURAL STONE retaining wall holds back steep slope; wall tilts slightly into plant covered hill.

ATTRACTIVE retaining wall of railroad ties set upright is separated from lawn by concrete strip.

RETAINING WALL between garden levels follows natural contours; poles set about 3 feet in the ground.

CAPPED WOOD retaining walls make terraced hillside vegetable garden; levels easy to reach by side steps.

SLOPING RETAINING WALL keeps driveway free of sliding mud; wall is veneer of flagstone over concrete.

TERRACE OF SMOOTH CONCRETE is supported by rubble stone wall topped with smooth coping for sitting.

WHITE SLUMP BLOCK walls retain steep slope, form elevated planting bed; draping plants soften wall.

WALL BUILT to provide pockets of soil where plants could find root room; most rocks came from the site.

SINGLE ROW of large rocks makes simple retaining wall to hold gentle, planted backyard slope.

How to Build Retaining Walls

An amateur builder can build a low retaining wall without too much difficulty. But a little knowledge about retaining walls can be a dangerous thing, according to many building inspectors. If you encounter any special retaining problems, don't attempt to handle the job alone. Seek professional help to insure that the wall is properly engineered and built. It's wise to seek such help any time you plan a retaining wall taller than 4 feet or are attempting to tame a hill with a greater than 36 percent slope. Call in an expert if you have to retain tricky soil such as marshy soil, adobe, or fill.

If you live in an area where earthquakes are a possibility, consult an expert to calculate possible earth slippage and retaining requirements for your wall.

Before you begin construction, regardless of the size of the wall, check your local building codes. Many cities require a building permit for any retaining wall, and some require that a retaining wall over 4 feet be designed and its construction supervised by a licensed engineer.

Keep yourself informed and don't take chances. Failure of a retaining wall could cause incalculable damage to your home or that of a neighbor down the hill from you if tons of earth

should come rolling down a hillside. Oozing mud caused by improper drainage provisions could also be a hazard.

Taming the slope

If your slope is not especially steep and the retaining wall does not have to be more than 2 or 3 feet high, you can build the wall yourself, using almost any one of the materials described above. Regardless of the type of wall you build, the slope of the hill will require some alteration before the wall can be installed.

Cutting and Filling. Since walls must be set on level ground, a slope must be cut and filled to create a plateau for each wall you build. The illustration (next column) shows three variations of the cut and fill technique.

In home subdivisions with sloping lots, people downslope may have to cut to gain a level area in their backyards. The people upslope may have to fill to gain level land space in front. Neighbors can often work together and move soil to their mutual advantage. Holding the hill is a project that affects everyone living on it.

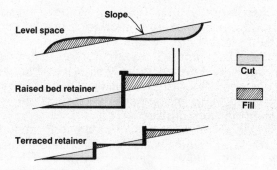

THREE WAYS to cut and fill: cut gentle slope above and fill below to achieve level ground; cut below and fill above behind retaining wall for raised bed. Cut and fill in stair-step fashion for terraced retainer.

Planting.

Once the walls are in place, the final step is usually planting the slope. (For gentle slopes or naturally rocky hills, planting alone may be enough to hold the soil.) The roots of the plants hold the soil while the outgrowth enhances the surface of the wall. Choose hardy, firm-rooted plants that cover well but don't spread too fast to control. Don't put off planting too long. Get the plants into the ground while the soil is still loose and easy to work; this allows roots to spread faster and hold the soil sooner.

Water control

The most crucial part of building a retaining wall is to plan for proper drainage. During the rainy season, the soil acts like a sponge, absorbing a large quantity of water. When the soil reaches a point of saturation, water begins to flow downhill below the surface. When this subsurface water reaches an obstruction, such as a wall, it collects and builds up pressure that may finally burst or undermine the wall. The wall builder must make ample provision to carry ground water through or around the wall.

Drainage. Getting rid of excess water that may take out your wall is sometimes a complex problem. Ditches, gutters, drain tile, and proper planting can be combined to divert the surface floods. Water flowing down the hill should be collected in gravel backfill and allowed to escape through weep holes in the wall or led away around the edge of the wall via drain pipe or drain tile. Any hillside drainage piping should funnel towards a storm sewer, ditch, or natural drainage where the concentration of water will not cause damage to property further down the hillside. Where weep holes allow water to drain out through the wall, a special surface gutter may be desirable to prevent the water from pouring over the lawn or

terrace. The gutter should be covered with a grating of wood or metal and should be pitched slightly to allow run-off.

A gutter built of planks will adequately handle drainage. Grating can be supported by a cleat on the upright plank and by a concrete shelf on the wall itself.

Clay tile of the half hexagonal type can be used but may be more likely to crack at the angles. Rounded gutters of concrete, spaced when the wall is built, will also serve effectively. Make all gutters shovel width for cleaning.

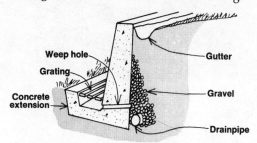

FOR DRAINAGE, gutter at top carries away surface water. Underground water behind wall escapes through drain pipe laid in gravel filled ditch or through small weep holes at base of retaining wall.

Holding the hill

After you choose a building material from the wide range of possibilities listed above, you're faced with the question of how to go about putting your wall together and how to place it. The technique and placement of your wall (or walls) depends largely on the material you plan to work with. Generally, it is better to use a series of low, close-set retaining walls than one high one. Low walls in a series are more pleasing to the eye, easier to plant with vines or shrubs, and less likely to lean or topple downhill.

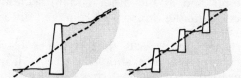

SERIES OF LOW, close-set retaining walls gives more strength in holding hill than single high wall.

Concrete cast in place. For a poured concrete wall, build solid forms that extend below the grade with an extending foot on the downhill side (see page 86). Before pouring the concrete, extend reinforcing rods or welded wire mesh down into the curved foot and to the top of the form. For pouring and finishing the wall, follow the directions given in the section on concrete walls.

Precast concrete posts. Reinforced internally with wire or rods, cast concrete posts make good retaining walls. They should be stacked in dovetail fashion, extending into the hillside and anchored with rods fastened to "deadmen" (sections of concrete buried in the slope). This type of retaining wall lends itself well to plantings.

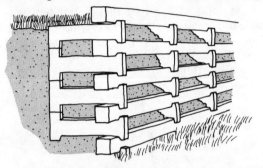

PRE-CAST REINFORCED concrete posts anchored with keyed posts run back into bank to retain slope.

Concrete blocks. Pour a solid concrete foundation and place vertical steel reinforcing rods in the concrete before it sets. Set the blocks in mortar over the reinforcing rods. Fill with grout the cells of the blocks containing the steel rods. For further construction details, see the chapter on building concrete block walls.

Bricks. A retaining wall built with bricks should be laid up in two tiers with steel reinforcing rods extending up from the poured concrete foundation between the two tiers. The space between the tiers should be filled with grout. For additional strength, tiers may be laid up in interlocking bonds or patterns. For details, see the chapter on brick walls.

Another method of laying up brick retaining walls is to stagger each course gradually inward. Use metal rods attached at one end in the mortar joints and at the other end in sections of poured concrete (deadmen) plugged into the slope of the hill.

These types of walls, though one of the most attractive possibilities, should not be built much higher than 4 feet because they simply do not tie together with the same strength as retaining walls made of other materials.

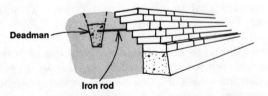

MORTARED BRICKS set on solid footing are staggered gradually. Steel rods attach to deadmen in slope.

Natural stone. Natural stone that has been cut and shaped can be laid up in mortar on a solid concrete foundation to form a sturdy retaining wall. Walls of cut stone are usually sturdier than brick because the finished stone wall is considerably wider than brick. Use an angle jig of boards to batter up the face of the wall (see the chapter on stone walls for construction details).

Uncut stones can be laid up dry to form quite sturdy low retaining walls that will hold low banks successfully. They can be laid up with earth pockets between them, then planted with vines or small, shrubby plants. Stones used should be laid so that they are pitched slightly inward against the thrust of the bank.

Stones may be laid up in a straight-faced wall or in the stepped back variety. The former has stones stacked one on top of the other with the larger stones concentrated on the bottom. The latter has stones stacked on the ground or dug in very slightly. The resulting wall has a massive and disorderly appearance but lends itself well to concealment with plants. See the chapter on stone walls for further details.

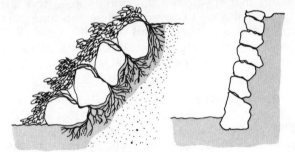

STACK LARGE STONES on ground of slope to form stepped-back retainer (left) or lay dry stone wall pitched slightly into bank; place large stones at bottom.

Wood. Retaining walls of wood should be built of redwood or cedar, both naturally resistant to rot and insect attacks. However, even these woods should be treated with preservatives before they are installed as retaining walls. Walls of 4 by 4-inch posts and planks should have each post braced into the hill as illustrated below. Railroad ties can be stacked and nailed together with long spikes or set upright and butted together in the

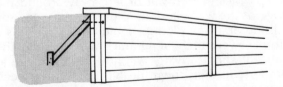

REDWOOD OR CEDAR wall built with 4 by 4s and planks. Brace each post securely into the hill.

slope or embedded in concrete. When stacked, railroad ties can be braced, like planks, with 4 by 4-inch posts. Or lay up ties in a staggered fashion to form a series of hillside steps.

Telephone or utility poles should be butted snugly in a fairly deep trench or may be set in a bed of concrete. Leave the poles at varying lengths to attain a casual, uneven look.

Natural bark-covered logs may be used for the walls of a raised bed and are available whole or split in half.

Raised beds and terraces

Retaining walls are much more than a grim necessity. Used in the form of raised beds, they become a valuable landscaping tool. Well designed raised beds have a strong architectural value. They also introduce into the garden interesting color and texture because of the range of materials used in their construction.

In a flat garden, raised beds bring a welcome relief to the landscape, providing an impressive display of plants and a smooth transition from one garden level to another. Raised beds can also be formed by large retaining walls holding back sizable hillsides. By back filling to the wall, you can obtain a large flat planting area. A high retaining wall holding an uphill slope provides privacy and, when properly planted, adds an attractive outside wall to a room.

Most commonly, raised beds are used in the form of four-sided elevated planting areas in the garden. Raised beds help bring plants up close; they are perfect for miniature or dwarfed plants that are often lost in the garden among taller-growing shrubs. When you bring them to eye level, you can really enjoy their diminutive charm.

Raised beds help make gardening easier by eliminating some of the stooping and bending from the garden chores. The built-up beds are easier to weed, cultivate, and water. Such plants as vegetables tend to produce earlier in a raised bed because the bed readily absorbs the sun's heat. Raised beds also help protect plants from running children and pets. If you're doing some special hybridizing, put the experimental plants in a raised bed where you can keep a closer eye on them.

In separating various elements of the garden plan, a raised bed can be very functional. With it you can separate a terrace from a parking strip or a lawn from a vegetable garden.

For plants that require good drainage, raised beds are the answer. Given proper moisture, the soil in the beds does not stay water-logged, as is sometimes the case with heavy, poorly drained surface soil.

Construction. Raised beds can be constructed with the same materials used for retaining walls as described above. The major difference in the construction of the raised beds is that they don't require the same degree of retaining strength as the larger wall that actually retains hillsides. Though some raised beds do retain hillside slopes, the variety that is raised from the garden floor bears very little soil pressure because the surface of the bed is widely spread. Most raised beds are no more than 12 inches high.

Tree walls

When the grade is lowered around a home site, a raised bed should be built to retain the soil around the tree at its original level. The wall does not have to be particularly strong, for its soil-retaining function is relatively minor. The wall should be reasonably water tight to prevent water from sluicing away from the tree roots.

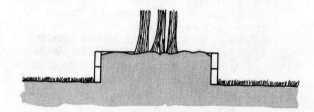

TREE WALLS retain raised ground around tree at original level; wall should be reasonably water tight.

Tree Wells

If the ground level is to be raised, a wall should be built around the tree before the grading is done. Soil is then piled up against it, making the wall the outside of a well. Old trees have lived for years with a certain balance of air, water, and nutrients, and it's difficult for them to adjust to sudden changes. Even a few inches of soil piled on top of the root zone seals off enough oxygen to harm the roots of the tree. The well allows the soil around the base and the roots of the tree to remain unchanged even though the surrounding grounds are elevated.

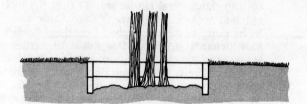

TREE WELL essential to tree life; keeps ground around tree at original level below surface of new grade.

Photographers

William Aplin: 26 (bottom left), 36 (left), 52 (top, bottom right), 70 (right), 80 (top right), 90 (center left). **Jerry Anson:** 79 (center left). **Morley Baer:** 9 (top left). **Robert Bander:** 72 (bottom left). **Nancy Bannick:** 73 (top left). **Ken Bates:** 28 (bottom left). **Clyde Childress:** 80 (bottom right). **Glenn Christiansen:** 19 (center left, bottom left), 44 (right), 47 (bottom left), 53 (top right), 64 (left), 71 (left). **Thomas Church:** 10 (top right, bottom right), 66 (bottom right). **Dearborn-Masser:** 27 (bottom left), 58 (bottom right), 59 (bottom right). **Lyn Davis:** 58 (bottom left). **Richard Dawson:** 17 (right). **Philip Fein:** 51 (top left, top right), 77 (left). **Richard Fish:** 25 (top right), 40 (right), 50 (right), 51 (right), 70 (left), 72 (bottom right), 84 (bottom right), 90 (top), 92 (left). **Frank L. Gaynor:** 76 (right). **Elizabeth Green:** 26 (center right), 36 (right). **Jeannette Grossman:** 25 (center). **Art Hupy:** 19 (bottom right), 24 (center right), 26 (top right), 91 (bottom right). **Tatsuo Ishimoto:** 47 (top right), 79 (bottom left). **Frank Jensen:** 8 (bottom left, top right), 53 (bottom right). **Lee Klein:** 76 (left). **Rene Klein:** 7 (bottom), 9 (bottom left), 10 (left), 23, 41 (bottom right), 47 (top left), 59 (top right), 63, 64 (bottom right), 65 (left), 73 (bottom left, right), 80 (left), 84 (left, top right), 85 (right), 90 (bottom left, bottom right). **Elsa Knoll:** 91 (bottom left). **Samson Knoll:** 37 (left), 50 (left). **Roy Krell:** 26 (top left), 53 (bottom left), 60 (top right), 66 (left), 71 (right). **Ells Marugg:** 5, 24 (bottom left), 41 (left, top right), 46 (left), 53 (top left), 64 (top right), 77 (right), 81 (bottom left), 92 (right). **Don Normark:** 17 (left), 25 (bottom), 28 (top left, bottom right), 46 (right), 52 (bottom left), 56, 58 (top right), 59 (top left, bottom left), 60 (top left, center left, bottom left, center right), 61 (top left, center left, bottom left), 65 (right), 72 (top left), 85 (top left), 91 (top left, top right). **Phil Palmer:** 24 (bottom right), 27 (top left), 85 (bottom left). **Maynard L. Parker:** 44 (left). **Charles Pearson:** 27 (center left). **Paul J. Peart:** 9 (top right). **Ray Piper:** 25 (top left). **Norman A. Plate:** 8 (center right), 78, 81 (top left). **Katherine L. Robertson:** 91 (center left). **Martha Rosman:** 19 (top right), 72 (top right). **Douglas Simmons:** 28 (top right). **Julius Shulman:** 42. **Donald W. Vandervort:** 16 (left, right). **Darrow M. Watt:** 7 (top), 8 (top left, bottom right), 9 (bottom right), 24 (top), 26 (bottom right), 40 (left), 58 (top left), 90 (center right). **Wenkam-Salbosa:** 61 (bottom right). **Steven C. Wilson:** 60 (bottom right), 61 (top right). **W. P. Woodcock:** 37 (right).

Designers

Guy Anderson: 46 (right). **Lorrin Andrade:** 25 (top left). **Baldwin, Erickson, and Peters:** 66 (left). **Bettler Baldwin** and **Owen Peters:** 40 (right). **Lyd M. Bond:** 26 (top left), 81 (right). **Roy Boyles:** 53 (top left). **Chaffee-Zumwalt:** 28 (bottom right). **Jack Chandler:** 70 (left). **Robert Chittock:** 28 (top left). **Thomas Church:** 10 (top and bottom right), 16 (bottom left), 19 (bottom left), 51 (top left), 66 (bottom right), 85 (right). **Eckbo, Royston, and Williams:** 9 (top left). **Ralph Edwards:** 8 (bottom left, top right). **Arthur W. Erfeldt:** 25 (center). **Ericksson, Peters, and Thomas:** 84 (bottom right). **John Fischer:** 60 (bottom left). **Mary Gordon:** 9 (bottom left), 10 (left), 23, 41 (bottom right), 65 (left). **C. Jacques Hakin** and **J. Charles Hoffman:** 91 (bottom left). **Lawrence Halprin:** 8 (bottom right). **Harris, Reed, Litzenberger:** 52 (bottom left). **Kenneth Hayashi:** 25 (top right). **Noble Hoggson:** 27 (center left). **Georg Hoy:** 37 (left). **Roland Hoyt:** 80 (top right). **Glen Hunt and Associates:** 61 (center left, bottom left). **Huntington** and **Roth:** 60 (top left). **Clayton James:** 85 (top left). **Jack Johnson:** 53 (bottom right). **William Louis Kapranos:** 7 (bottom), 47 (top left), 59 (top right), 63, 64 (bottom right), 73 (bottom left, right), 84 (top left, right), 90 (bottom left, bottom right). **Kenneth Kelly:** 27 (bottom left). **Warren E. Lauesen:** 47 (bottom right), 60 (top right). **Neal Lindstrom:** 41 (top right). **Frank E. Martin:** 71 (right). **Cliff May:** 37 (right), 42. **Mitchell McArthur Gardner O'Kane Associates:** 25 (bottom). **Roland Molen:** 53 (bottom left). **Tom Nishamura:** 19 (bottom right). **Courtland Paul:** 9 (top right). **Marshall W. Perrow:** 24 (center left). **Burr Richards:** (65 right). **Dan Rolfs:** 19 (center left). **Gil Rovianek:** 24 (bottom left). **Royston, Hanamoto, Mayes,** and **Beach:** 24 (bottom right). **J. R. Schibbley:** 76 (right). **Mildred L. Schuyler:** 41 (bottom left). **Anthony Silvers:** 27 (bottom right). **Jack C. Stafford:** 92 (right). **Kathryn Imlay Stedman:** 8 (top left). **Kay Scott:** 50 (left). **William Talley:** 91 (top left). **Maria Wilkes:** 79 (lower right). **Kenneth Wormhoudt:** 51 (right). Cover brickwork designed by **Roger Flanagan.**